AF560060

Insect Classification and Systematics

NIPA® GENX ELECTRONIC RESOURCES & SOLUTIONS P. LTD.
New Delhi-110 034

About the Author

Prof. Neerja Agrawal graduated in bio stream and obtained Masters in Zoology with specialization in Entomology, holding meritorious position in University of Allahabad. She won State Fellowship to pursue her Ph. D. in Entomology from Chandra Shekhar Azad University of Agriculture and Technology (CSAU), Kanpur, under the guidance of Professor Entomology & Dean Agriculture Dr. K.D. Upadhyay. She completed International Course on 'Applied Taxonomy of insects and mites of agricultural importance' at CAB International Institute of Entomology, London in 1986 through CFTC Fellowship (U.K.).

Having started her research and teaching career in 1987 as Junior Scientist/ Assistant Professor in Deptt. of Entomology, C.S.A.U.A.&T., Kanpur, she retired as Dean and Professor & Head in 2018. After retirement she was selected by ICAR to join as Emeritus Professor at CSAU, Kanpur. She has been teaching Insect morphology, advanced insect systematics and insect ecology and guided 22 M.Sc. (Ag), 10 Ph.D. researchers and 1 Post- Doctoral Fellow. She led three National Projects as PI and two Projects as Co-PI., sponsored by Ministry of Agriculture, Govt. of India and ICAR, respectively. She presented Country Report in the 'First Int. Symposium on Fruit flies in the Tropics' held in Malaysia (1988). Later, she delivered Research Papers in many National/International Conferences including Malaysia (1990), Japan (1993), Indonesia (2018) and France (2019). She has contributed more than 80 research papers in the Journals of National and International repute, 5 books, 4 practical manuals and 10 book chapters, besides being actively engaged in print/ electronic media and website tutorials for University students. Her website www.profneerja@wordpress.com caters instructional material for different PG courses in Entomology.

Prof. Neerja had addressed Women's Science Congress in the 100th Indian Science Congress (ISC), Kolkata in 2013, as Guest Speaker from academic sector and was elected Recorder in the section of Agriculture and Forestry Sciences for the years 2017 and 2018 at ISC. She is Fellow Entomological Society of India, member editorial board and life member of many scientific societies. She is recipient of Smt. Rewati Singh Memorial Award- 2018 for outstanding contribution in the field of Science and Agriculture by Society for Recent Development in Agriculture; Dr. A.P.J. Abdul Kalam Education Excellence Award -2019 by Indian Solidarity Council and Lifetime Achievement Award-2019 in Entomology by Agriculture & Environmental Technology Development Society.

Insect Classification and Systematics

Professor Neerja Agrawal F.E.S.I.
Ex. Emeritus Professor (ICAR)
Professor and Head (Retd.)
Department of Entomology
C.S. Azad University of Agriculture & Technology
Kanpur, Uttar Pradesh

NIPA® GENX ELECTRONIC RESOURCES & SOLUTIONS P. LTD.
New Delhi-110 034

NIPA® GENX ELECTRONIC RESOURCES & SOLUTIONS P. LTD.

101,103, Vikas Surya Plaza, CU Block
L.S.C. Market, Pitam Pura, New Delhi-110 034
Ph : +91-11-43860225, Mob.: +91 9717133558, 9540816132
E-mail: newindiapublishingagency@gmail.com
Website: www.nipaersources.com

Print ISBN: 978-93-95763-79-0
ebook ISBN: 978-93-58870-87-9

Composed and Designed by NIPA®.

Preface

I don't care how small or big they are, insects freak me out.
–Alexander Wang

Hexapods or insects evolved from many legged segmented soft bodied Onychophora like organisms through a process of continuous cephalization i.e concentration of sense organs, nervous control etc. at the anterior end of the body, forming a head and brain. They evolved much earlier than man, around 350 million years ago while the existence of man dates back to 1.5 million years. Insects are continually evolving even today. The diversity of form and habitat results in presence of insects everywhere in air, water and land. The insects coming under Phylum Arthropoda, share some common characters with other arthropodan classes. The major advancement in insects were the origin of wings that facilitated dispersal, helped them to occupy new niche and being first flying organisms, it reduced their enemies in air and increased their chances of survival. Insects impact the entire food chain, food web, have economic importance and the fossil record of insects tells us much about other life on earth.

System of insect classification is based on their evolutionary history. Insect fossils were abundant in the Silurian and Devonian. In the Carboniferous and Permian periods evolution of winged insects occurred with hemimetabolous type of metamorphosis. The development of wing and pupal stage causes a profound effect on the success of insect world, diversity and survival. Since the last 50 years, systematics had gone through a remarkable renaissance. This is because the study of organic diversity is a major integral part of biology. With the use of computers and concurrent attempt to automate classification through electronic data, the science of classification has increased multifold. The increasing need of applied aspect of taxonomy such as the correct identification and classification of species in agriculture, public health, ecology, genetics and behavioural entomology has made insect systematics more popular.

Working on this book have been largely inspired by the International Course on "Applied Taxonomy of insects and mites of Agricultural Importance" which I completed at Commonwealth Institute of Entomology (CAB International Institute of Entomology), London U.K. The knowledge gained during training at CIE, had added high standards in my academic career and gave me an insight to write a book. The aspiration for a book on insect systematics was seeded some 17 years back when I went through a notification by ICAR, inviting University level books in Agriculture and allied subjects by Indian authors. Somehow it could not be materialized that time and I continued teaching and absorbing knowledge. Experience of teaching has helped a lot in preparing the chapters.

Over a period of time, I got a chance to go through several courses / trainings on taxonomic procedures of various insects including parasitoids, fruit fly and bioagents at Project Directorate of Biological Control, Bengaluru (now National Bureau of Agricultural Resources), Indian Institute of Horticultural Research, Bengaluru and Directorate of Oilseed Research, Hyderabad. This has made the circle of literature wider, encompassing new learning and experiences to be incorporated in the book.

Systematics is the mother science to all disciplines of biological sciences. It includes taxonomy, theory and practice of classifying organisms, and also attempts to find any and all relationships to unravel very process of species diversification. In light of the present scenario of changing biodiversity, new methods of nomenclature are emerging in the field of biosystematics which have been included in the text. The immense advancement in theoretical and applied aspects that have taken place in the past decade, implied the author to bring out a completely enlarged version of the present book. The book attempts to present to students a detailed yet reasonable knowledge of insect classification and systematics.

The idea behind writing this book is that a reference book is very much required, written in simple language covering all the topics related to insect classification and systematics. I have selected materials primarily from the point of view of students and have chosen information from my own practical knowledge that is helpful in the proper identification of insects such as in chapter "Procedures in Identification". In an endeavour to bring together the more important elements of modern taxonomic theory and practice, a critical evaluation of taxonomic theory has been discussed in detail with "Modern Classification of insects", "History of Classification" and "History of insect Systematics". "Taxonomic collections, Curation and Preservation," have been extensively explained and every attempt has been made to make understand the importance of different steps required in insect collection. While carrying out the work on insect collection, pinning and labelling, this book will prove a working guide to many students perusing higher education in the field of Entomology. Much emphasis has been laid on numerous illustrations and diagrams pertaining to their clarity and ease to grasp.

The objective is not just to make a contribution to theoretical knowledge but to establish a sound foundation for practical operations. A glossary to understand exactly the meaning of such terms as species, taxon, category, classification, type etc. and many entomological terms have been incorporated as these are important for the entomologist at every stage of their work. In the end, "Expected questions" will lend a hand to assist students of all Universities /Agriculture Universities, in their studies and to prepare for a suitable career in agriculture sector. I am afraid that errors will reamin, despite all my efforts. However, transcending the boundaries of this strive, I look forward to criticism and constructive suggestions for the improvement of the book.

I got support in various ways, such as suggestions, encouragements and blessings of different personals who have been my colleagues, learned teachers, research personals at different parts of the country, friends and well-wishers. I would like to express my deep sense of gratitude to those noble authors whose published work, reviews and research work had helped me to gather information for relevant topics. The role of digital resources is no less important. I deeply appreciate the help and encouragement I have had from my two daughters, who continue to maintain an undiminished interest in this book. Ar. Charu Priya worked tirelessly side by side, connecting with me on internet, in drawing the diagrams and illustrations, while Dr. Kritika Agrawal contributed her skills of composing and editing.

In the last but not the least, I am thankful to my publisher NIPA® and their amazing team, who worked patiently as many a times I was travelling and therefore, it took longer to complete the manuscript.

Neerja Agrawal

Contents

1

Origin of Arthropods

Arthropod life began very early in Palaeozoic region, about 600 million years ago in the sea. The arthropod foundations were laid with the evolution of metamerism in segmented worms, and in the ancestral arthropods it resembled the now extinct Trilobites. They had harder cuticle forming the exoskeleton and a pair of jointed legs with each of its metameric segments.

The Trilobite stock soon diversified into:

a) Chelicerates.

b) Mandibulates.

a) Chelicerates: In Chelicerates, the anterior body segments fused to form the prosoma, bearing a pair of chelicerae, the feeding organs; a pair of pedipalps and four pairs of walking legs. The remaining segments formed the legless opisthosoma. This gave rise to Eurypterids (Extinct) Xiphosurians, Pycnogonids and the highly successful Arachnids, including the scorpions, spiders, ticks and mites.

b) Mandibulates: Body segments of Mandibulates consolidated and gave rise to three major classes- Crustacea, Myriapoda and Insecta.

1. **Crustacea:** The anterior body segment formed the cephalothorax bearing two pairs of antennae, a pair of mandibles, two pairs of maxillae and five pairs of legs. The remaining segments formed the abdomen bearing variable number of legs and a terminal telson. Most of them never left the ocean and became important part of food chain to sustain larger vertebrates including men from the ocean ecosystem. They include shrimps, prawns, cray fishes, lobsters, crabs etc.

2. **Myriapoda:** For the first time in arthropod evolution, the anterior body segments fused to form the definite head capsule bearing a pair of antennae, a pair of mandibles and two pairs of maxillae. The remaining segments formed the trunk region bearing at least a pair of legs on each body segment. These animals were close to insects but had too many legs to coordinate their body movements. They had left the ocean and became

the first to take to terrestrial habitat where green plants were just beginning to grow. These animals gave rise to herbivorous Diplopods, the millipedes and the carnivorous Chilopods, the centipedes.

3. **Insecta:** With the adaptation to terrestrial habitats, some of these animals gave rise to insects. They had the first six anterior body segments fused to form the head, the next three formed the thorax and the remaining eleven segments formed the abdomen. The insect head became specialised for feeding and sensory perception with one pair of antennae, one pair of mandibles, maxillae and labium. The thorax became the locomotion centre and carried three pairs of legs and two pairs of wings in winged insects. The abdomen was without legs and became specialised for reproduction.

Insects are highly specialized group of invertebrates belonging to the largest animal Phyla, the ARTHROPODA, which also includes in it, the prawns, crabs lobsters, centipedes, millipedes, scorpions, ticks, mites and spiders. Practically three fourth of the animal species come under this phylum.

Arthropods are

1. Bilaterally symmetrical animals with their bodies divided into a number of rings of segments and covered by an exoskeleton made up of a substance called chitin.
2. They have jointed appendages provided with independently movable muscles and inserted to special pockets of the exoskeleton.
3. Body cavity known as haemocoel in which blood is circulated. All internal organs are bathed in it.
4. Presence of ventral nerve cord.
5. Special organs for respiration known as tracheal system which are branching of tube like trachea.
6. Ecdysis or moulting is a characteristic of all Arthropods where by cuticle is shed off at regular intervals in order to accommodate the growing tissues.

Phylum is divided into

1. Crustacea (Prawns and crabs).
2. Onychophora (Peripatus).
3. Myriapoda (Millipedes).
4. Diplopoda (Centipedes).
5. Insecta or Hexapoda (Insects).

6. Arachnida (Spiders, scorpions, ticks and mites).
7. Trilobita (extinct group).

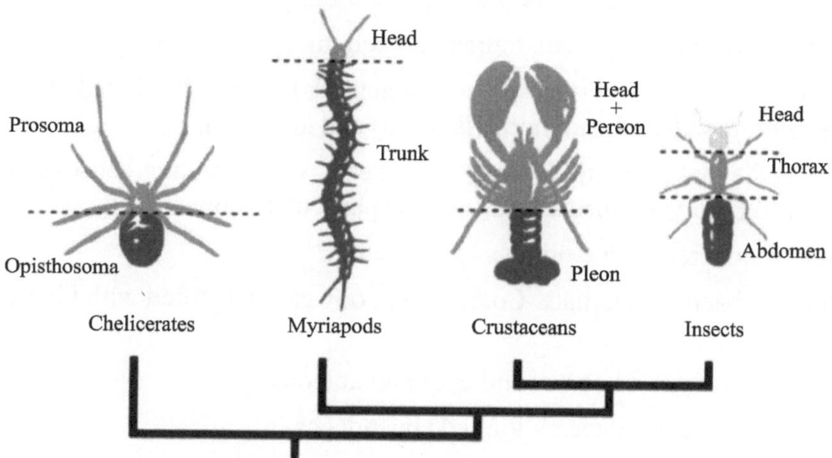

Figure 1.1. Origin of arthropods

Phylum Arthropoda

Arthropods were first studied by **Aristotle**. It is the largest phylum in the animal kingdom. **Von Siebold** coined the term Arthropoda. It is derived from the Greek word *Arthos* = jointed; *podas* = legs. It constitutes the largest Phylum of animal Kingdom. At least 80 per cent of all known species of animals are arthropods. This phylum comprises invertebrate animals with the following characters:

1. They are bilaterally symmetrical, triploblastic and metamerically segmented.
2. Jointed legs with varied functions. Body is covered by hard chitinous exoskeleton.
3. Body divided into head, thorax and abdomen or cephalothorax and abdomen.

General Characters

1. Cosmopolitan in distribution, found in aquatic, terrestrial and aerial forms. Some are ecto-parasitic and vectors of diseases.
2. Body have jointed appendages or legs (which are modified to different structures to perform different functions like jaws, gills, walking legs, paddle). There may be 3 pairs, 4 pairs, 5 pairs or many pairs.

3. Body is triploblastic.
4. Bilaterally symmetrical.
5. Organ systems with high level of organization.
6. Body is divisible into head, thorax and abdomen.

   Note: In some (crustacean and arachnids) body is divisible into cephalothorax (head and thorax is fused) and abdomen.
7. This is the first group to develop a true head, which contains sense organs and feeding organs specialized for their particular habitats.
8. Body is covered with chitinous exoskeleton.
9. They are haemocoelomate. Coelom i.e. body cavity is filled with blood or fluid.
10. Head bears a pair of compound eyes and antenna.
11. Locomotion takes place by jointed appendages.
12. Digestive system is complete, straight and well developed.
13. The mouth bears mouth parts for ingestion of foods. Mouths are modified for chewing, biting, sponging, piercing, siphoning and lapping.
14. Respiration takes place by general body surface or gills (crustaceans) or trachea (insects, Diplopoda and Chilopoda) or booklungs (Arachnida) and book gills (horseshoe crabs).
15. Circulatory system is of open type i.e. do not have blood vessels and enters directly into the body chambers. The blood is colourless.
16. Excretion takes place through Malphigian tubules (in terrestrial form) or green glands or coxal glands (in aquatic forms).

    NOTE: Aquatic forms are ammonotelic, terrestrial forms are uricotelic.
17. Nervous system is of annelidian type, which consists of brain and ventral nerve cord.
18. Unisexual i.e. sexes are separate.
19. Fertilization is internal or external.
20. They are either oviparous or ovoviviparous.
21. Development may be direct or indirect.
22. Sensory organs include antennae, sensory hairs for touch and chemoreceptor, simple and compound eyes, auditory organs (in insects) and statocysts (in crustacean).

Phylum Arthropoda is divided in the following classes:

Class 1 - Onychophora (claw bearing)

1. Terrestrial in habit.
2. Body not divided into distinct regions.
3. One pair of antenna, many pairs of unjointed legs.
4. Breathing through trachea.
5. Example- *Peripatus.*

Class 2 - Arachnida (Arachne)

1. Body divided into cephalothoraxes (Prosoma) and abdomen (opisthosoma).
2. Prosomatic appendages are one pair of chelicerae and one pair of pedipalpi.
3. Four pairs of walking legs.
4. True jaws are absent.
5. Terrestrial or aquatic in habit.
6. Example- Scorpion, spiders, ticks and mites.

Class 3 - Crustacea (Crust - shell)

1. Aquatic in habit.
2. Body divides into cephalothoraxes and abdomen.
3. Two pairs of antennae and five pairs of walking legs.
4. Breathing through gills.
5. Example – Crab, Prawns.

Class 4 - Chilopoda (Chilo- lip; poda - appendages)

1. Terrestrial in habit.
2. Body divided into head and trunk region.
3. Head with one pair of antennae, one pair of mandible and two pairs of maxillae.
4. Each trunk segment with one pair of legs.
5. The 1st pair of leg is modified into poison claw.
6. Example- Centipede *(Scolopendra*).

Class 5 - Diplopoda (Diplo-two; poda- appendages)

1. Terrestrial in habit, body divided in to head thorax and abdomen.
2. Body elongated, cylindrical and vermiform.
3. Head five segmented and bears one pair of antenna, one pair of mandible and a long maxilla.
4. Thorax four segmented and each segment with one pair of walking legs.
5. Each abdominal segment carries two pairs of walking legs.
6. Breathing through trachea.
7. Example – Millipede (*Julus).*

Class 6 - Trilobita (an extinct group)

Class 7 - Insecta or Hexapoda (Hexa- six; poda- appendages)

1. Terrestrial or aquatic in habit.
2. Body divided into head thorax and abdomen.
3. Head carries one pair of antenna, one pair of mandible and one pair of maxilla.
4. Thorax three segmented (Prothorax, mesothorax and metathorax)
5. Each segment carries one pair of legs and meso and meta thorax also carry one or two pairs of wings.
6. Breathing takes place through trachea.
7. Example – Grasshoppers, locusts, crickets, termites, moths, Butterflies beetles, plant bugs etc.

2

Systematics and Taxonomy: Relationship between Systematics and Taxonomy

Origin of Systematics

In Vedas (1500 BC), Ramayana (1900 BC), Mahabharata (1400 BC), and Upanishad (350 BC) mention had been made about various animals. The number of known species of animals is much greater than that of plants and has been estimated at about more than one million. When subspecies are included, the number may double to more than two million named forms of animals. The new animals are being described at the rate of about 10,000 per year. Metcalf (1940) estimated that one and a half million species names are already applied while Silvestre (1929) calculates an estimated 3 million probable species of insects.

Systematics

Systematics is a term coined by a Swedish naturalist Linnaeus (*Systema Naturae* 1735) which is a Latinized Greek word *Systema*, as applied to the systems of classification. According to Simpson, "Systematics is the scientific study of the kinds and diversity of organisms and any or all relationships among them." It includes Taxonomy, identification, classification and nomenclature. Now-a-days, both terms are used inter-changeably in the fields of plant and animal classification.

Taxonomy

The name taxonomy was first proposed by de Candolle (1813) for the classification of plants. The term taxonomy is derived from a Greek word *Taxis* means arrangement and *nomy* means law. Thus, arranging the living organisms according to certain law is taxonomy.

"Taxonomy is theoretical study of classification, including its basis, principles procedures and rules". It is the science of classification of any living organism. Taxonomy is based on basic fields of morphology, physiology, ecology and genetics. It is an integration of many kinds of knowledge theory and method applied to the field of classification. Classification is the first step to obtain any kind of biological knowledge, in an orderly system. Thus, naming of living organism, its description and classification of all plants and animals is necessary.

Levels of Taxonomy

The Taxonomy of a given group passes through several stages. These are:

a) **Alfa Taxonomy:** It emphasises on description of new species and their preliminary arrangement in comprehensive genera. It is the level at which the species are characterised and named.

b) **Beta Taxonomy:** Species are grouped in higher categories based on their phylogenetic relationships and emphasis is placed on development of a sound classification. It refers to arranging of species into a natural system of lesser and higher categories.

c) **Gamma Taxonomy:** In this, much attention is paid to infra specific variation, to various sorts of evolutionary studies, and interpretation of organic diversity. It refers to analysis of intraspecific variation and to evolutionary studies.

New Systematics

Current taxonomy is customarily referred to as new systematics. The purely morphological species definition has been replaced by a biological definition which takes ecological, geographical, genetic and other factors into consideration. Nomenclatural problems occupy a subordinate position in systematic work. This seems different from simple Taxonomy of Linnaeus or Fabricius and new terms have been suggested for the new species. Although modern taxonomy may be referred to as new systematics or as experimental taxonomy, it would be misleading to use these terms in contradiction to taxonomy. The modern systematics is showing an increasing interest in the formulation of generalization, for which naming and describing of species is only the first step. Thus, principal tasks of systematics are- **Identification** (Analytical stage) and **Classification** (Synthetic stage).

Functions of Systematics or Taxonomy

- Organize our knowledge about living organisms.
- Provide means of communication.
- Provide frame work for information, storage and retrieval system of great capacity.
- Help in developing principles of predictive value.

Relationship between Systematics and Taxonomy

Taxonomy and systematics go hand in hand. They are two concepts related to study of diversification of living things and the relationships among them. The main difference between taxonomy and systematics is that taxonomy is involved in classification and naming of organisms whereas systematics is involved in determination of evolutionary relationships of organisms.

This means that systematics ascertains the sharing of the common ancestry by different organisms.

In taxonomy different organisms are scientifically named and grouped in different taxonomic levels. Organisms are grouped based on their evolutionary relationships. Therefore, taxonomy is considered as branch of systematics. Both taxonomy and systematics use morphological, behavioural, genetics and biochemical observations.

Difference between Systematics and Taxonomy

S.No.	Systematics	Taxonomy
1	Systematics is the study of the kinds and diversity of organisms and the relationships among them.	Taxonomy however, is the theory and practice of identifying organisms, describing, naming and classifying organisms.
2	Systematics manifests the common ancestry by different organisms.	Taxonomy can be considered as a branch of systematics

3

Relationship of Insects with Other Classes of Phylum Arthropoda

Arthropods occur up to the height of over 6,080 meters on land and up to the depth of about 59,728 meters in water. Different arthropod species are adapted for living in air, on land, in soil and in fresh, brackish or salt water.

Many insects are economically important for being pests of crops, stored grains, stored food, household goods and clothing etc. Some insects and arachnids are in one way or the other responsible for certain diseases of crops, domestic animals and human beings. Some arthropods, on the other hand, such as crabs, lobsters, prawns and shrimps form a part of human diet while, small sized crustaceans are staple food for fishes. Many parts of the world have witnessed some insects and spiders eaten by many land animals including man.

Insects share many characters with other arthropods, but the other arthropods have not been able to exploit these to the extent insects have evolved. The presence of functional wings is a very important character which insects do not share with other arthropods, that is why insects have distinct advantage over other arthropods in their struggle for existence and dominance. A comparative account of Insecta and other arthropod classes is given in Table 1.

Table 1: Relationship of Insecta with other Classes of Phylum Arthropoda

S. No.	Characters 1	Onychophora 2	Arachnida 3	Crustatia 4	Chilopoda 5	Diplopoda 6	Insecta 7
1.	**Economic importance**	No economic importance. They are of academic interest, known as connecting link between Annelida and Arthropoda.	Pests, predators and disease carriers. Scorpions are predatory on flies, cockroaches, crickets etc. Mites are pest or crops or predatory. Ticks are parasitic on man or animals. Spiders are predatory.	Important sea food in human diet such as prawns, lobsters, crabs etc. Small crustaceans serve as food for fishes. Certain crabs are pests of paddy, coconut palm.	Predatory, prey upon insects, earthworms, lizards and sometimes even mite.	Feed on dead plants, animals. They are pest of living plants, damaging underground parts of certain plants of economic importance. Some secrete phenol or produce hydrogen cyanide.	Insects are pests of plants, by sucking plant sap, biting and chewing, rolling of leaves, making galls, boring and tunnelling different plant part or pest of stored products. Some are poisonous, others vectors of plant diseases. Some are beneficial like honey bee, lac insect, silk worm and help in pollination.
2.	**Habitat**	Terrestrial	Terrestrial	Aquatic and few terrestrial	Terrestrial	Terrestrial	Many terrestrial, very few aquatic
3.	**Size and shape**	Caterpillar like, cylindrical. The largest.	Varied shapes in different orders, body covered by chitinous cuticle.	Microscopic to big crabs, the largest is spiny lobster. Shape varies, body covered by chitinous cuticle which becomes thicker where no movement is required.	Slender elongate, segmented forms, flattened dorsoventrally. Some tropical species are 15-20 cm long. The South American sp is 26.5 cm long.	Cylindrical body of many segments. The African sp is 28 cm long.	Mostly insects are 2-40 mm long. Some Hemipterans, Coleopterans are larger and walking stick, a Phasmid, is upto 33 cm long.

4.	**Body regions**	Not distinct	Three- Pro, meso and metasoma eg. Scorpion Two- Pro and Opisthosoma eg. Spider.	Two- cephalothorax and abdomen.	Two- Head and multisegmented trunk.	Two- Head and multisegmented trunk.	Three- Head, thorax and abdomen.
5.	**Antenna**	1 pair	No antenna	2 pair- antinule and antenna	One pair	One pair	One pair
6.	**Visual organs**	One pair of simple eyes	One pair- simple eyes	One pair- stalked compound eyes	Single or cluster of simple eyes	Two clumps of many simple eyes	Simple and compound eyes (one pair)
7.	**Locomotory organs**	Many pairs of unjointed legs.	Four pairs of legs on cephalothorax in adult stage, in some larval forms only three pairs.	Minimum five pairs of biramous legs.	One pair of leg per segment (First pair of legs modified as poison claws).	Two pairs per segment (No poison claws).	Three pairs of legs on three thoracic segments and two pairs of wings on meso and meta thorax.
8.	**Mouth parts**	Mouth rimmed by a fleshy fold containing 2 small horny jaws and 2 blunt oral papillae.	1 pair of claw-like chelicerae, 1 pair of pedipalpi short, leg like with enlarged bases forming maxillae for squeezing and chewing food.	1 pair of mandibles, 2 pairs of maxillae and 2 pairs of maxillipedes.	1 pair of mandibles and 2 pairs of maxillae.	Mandibles 1 pair and maxillae 1 pair.	1 pair of mandibles, 2 pairs of maxillae and a labium. Mouth parts adapted for chewing, sucking or lapping.
9.	**Respiration**	Cutaneous	Book lungs (Scorpian) and tracheal (spider)	Gill breathing	Tracheal	Tracheal	Tracheal
10.	**Blood fluid**	Haemolymph	Haemolymph	Haemolymph	Haemolymph	Haemolymph	Haemolymph

11.	**Circulatory system**	Heart with ostia	Heart with ostia	Heart with ostia	Heart with ostia	Heart with ostia	Heart with ostia
12.	**Reproduction**	Sexes separate, reproductive organs paired but with single external opening.	Sexes separate, female larger than male.	Sexes separate except in some parasitic forms, sex openings paired.	Sexes separate, having one dorsal gonad. Ventral genital opening near the posterior end.	Sexes separate, female gonopore between the coxae of the fourth pair of legs. In males opening on sternite four.	Sexes separate, female larger than male. Genital opening at the end of the abdomen.
13.	**Development**	Anamorphosis	Metamorphosis absent	Anamorphosis	Metamorphosis	Metamorphosis	Metamorphosis
14.	**Habit**	Feed on organic matter	Phytophagous and predacious	Herbivorous and carnivorous	Carnivorous	Herbivorous	Phytophagous, Predators and parasitoids
15.	**Special features**	Link between Annelida and Arthropoda.	Life cycle- Egg, larva, nymph, adult with 3 pairs of legs.	Calcification strengthens exoskeleton.	Opisthogenital gonopore present in the terminal segment and nymph with 4 pairs of legs.	Progogenital gonopore in 3rd segmen.	Genital structures on 8th and 9th abdominal segments. Brain with proto, deuto and tritocerebrum.

4

Origin of Insects: Dominance of Insects in Animal Kingdom

Origin of Insects

Insects are highly specialized group of invertebrates belonging to the largest animal Phyla, the ARTHROPODA, which also includes in it, the prawns, crabs, lobsters, centipedes, millipedes, scorpions, ticks, mites and spiders. Practically three fourth of the animal species come under this phylum.

The Arthropods evolved during Cambrian period from primitive annelidan ancestor. The main evolutionary stem of Arthropod is represented by the Crustacia- Insecta- Myriapoda.

The insects originated in the Devonian period of Palaeozoic era. Diplura and Thysanura are the most primitive wingless insects while mayflies, damselflies, dragonflies and stoneflies represent the primitive winged insects.

Evolution

Evolution is a process whereby population are altered over a period of time and may split into separate branches, hybridize together or terminate by extinction.

The Evolutionary History of Insects

The origin of insects dates back to about 350 million years presumably in the lower Devonian period or before which the fossils are less or practically none. The numbers of the aerial and land forms were very few. On the basis of Arthropod morphology, myriapods appear to be closely related to insects. Of the myriapods, symphyla share more characters in common with the primitive insects and it would be suggested that myriapods and insects arose from common stock or from myriapod like ancestor. Evidence of this is seen from the fact that:

- The first instar of many myriapod diplopods show many features in common with insects, such as a head with 6 segments, a thorax of 3 segments, each with a pair of legs and an abdomen of about 5 segments, without appendages or with vestigial legs.

- The fact that among apterygotes leg like appendages are present on the abdominal segments and that the collembolans have a short abdomen of 6 segments.
- The presence of a pair of antennae with variable number of segments, each supplied with muscles, is a feature common to all myriapods and some primitive insects like Diplura and Collembola.
- Also important are the similarity of malpighian tubules and tracheal tubes and the fact that ecdysis occurs through a transverse slit on the hind border of the head in Myriapoda, Protura and Collembola.
- An epicranial suture or ecdysial suture is common to many symphyla and insects.
- The structure of post mandibular appendages is similar as also the nature of hypopharynx.
- The presence of styles and eversible sacs are common to Symphyla and Diplura.
- The terminal cerci of Symphyla appear to correspond to those of insects.
- The development of Symphyla shows that they also have a trunk composed of 14 segments as in insects.
- In spite of all these common traits, symphylids are progoneates with the reproductive system opening at the anterior end.
- Many myriapods have a peculiar hexapod larva, which suggests that the insect could have been originated from myriapod like ancestors by a process of neoteny, i.e. young one showing precocity.

For many decades, it was thought that insects branched off of millipedes, which are known to have colonized the land as early as 428 million years ago, during the **Silurian** Period. But recent genetic studies suggest that insects more likely split off from crustaceans sometime around 410 million years ago. (Fig. 4.1).

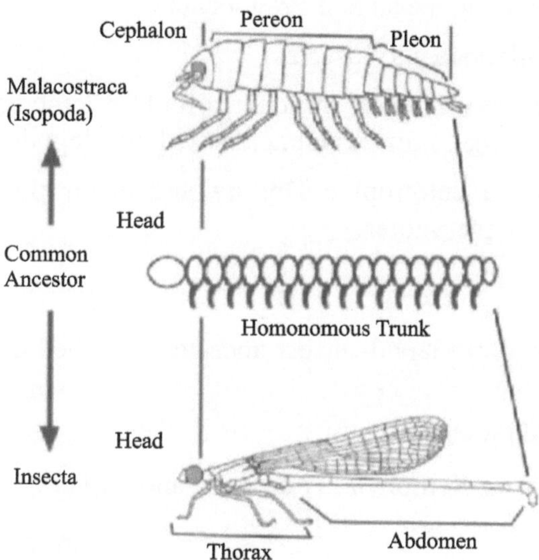

Figure 4.1 Origin of insects

Theories of Insect Evolution

1. Handlirsch's Theory

- Handlirsch's Trilobite Theory suggests that the insects have developed from marine ancestors, the trilobite.
- The wings of insects according to his view, would be homologous to the paranotal lobes i.e. the broad lateral extensions of the sides of the segments of trilobites.
- The trilobites also have a single pair of antenna, compound eyes and three ocelli, as in insects.
- The theory has no standing today because it would mean that the winged insects arose earlier than the apterygotes.

2. Hansen's Theory

Hansen proposed that insects originated from crustacean group Syncarida.

3. Crampton's Theory

Crampton advocated that the insects originated from Isopods.

4. Tillyard's Theory

- He believed that both the Myriapods and insects have descended from a common hypothetical ancestor which he called as Protapera.

- This broke off in two lines; Opisthogoneata and Progoneata.
- The former again split in to Chilopoda and Insecta.
- From the insectan branch there was very early branching off of Collembola and Protura, both the groups having common characters with myriopods.
- This stem then divided into the entotrophic Diplura and ectotrophic Thysanura, from which arose the pterygotes.

5. Ross's Theory

- Ross believed that the terrestrial myriapod- insect ancestor branched off into 3 lines of descent.
- One representing the Class Chilopoda.
- Second representing three Classes; Symphyla, Pauropoda and Diplopoda.
- Third representing Class Insecta.

The first two lines of descent i.e. Chilopoda, Symphyla, Pauropoda and Diplopoda possessed legs on almost all the body segments. In further evolution, the legs on three segments, posterior to gnathocephalon were retained for locomotion while locomotory appendages on rest of the body segments were lost. The three distinct body regions, head, thorax and abdomen became clearly distinct. These evolutionary modifications confirm the general belief that insects have evolved from a myriapod or protomyriapod like ancestor.

6. Menton's Theory

- Menton considered that one of the polyphyletic lines of evolution is Phylum Uniramia which included Onychophora, Myriapoda and Insecta.
- She thought that the ancestor of Phylum Uniramia were soft bodied with many pairs of unsegmented lobopodial appendages.
- The Onychophora retained these characteristics in addition to gaining the simple sclerotized jaws on the second cephalic segment.
- The ancestors of Myriapod- Hexapod line of evolution had mandible evolved on fourth segment. This line then branched off into two directions – one giving rise to Myriapoda (segmented jaws and many pairs of leg) and the other giving rise to Hexapoda (unsegmented jaws and legs moving in different way).

Dominance of Insects in Animal Kingdom

Factors that Influence Dominance

1. More number of species

Insects are about 85 per cent in the animal kingdom. Total number described so far is about 1 million.

2. Various habitat

Insects are found in almost all habitats; whether in soil, water or snow thriving in all varied conditions.

3. High fecundity

A single species may reproduce in large numbers thereby large population of insects surviving on earth. For example, locust's swarms comprise 10^9 number of individuals.

4. Long geological history

Insects were evolved some 350 million years ago and during the course of time they have gone through great variety of adaptations to cope up with different conditions.

5. Capacity for flight

Since they possess wings, which are the lateral expansion of exoskeleton, the adaptation helped them to use flight for the following purpose:

i. To seek food, mate, shelter and oviposition sites.

ii. To colonize in a new habitat and also to exchange habitat.

iii. To escape from enemies and unfavourable conditions.

iv. To migrate (i.e. for long distance travel e.g. locusts).

6. Wide range of adaptability

Insects are the earliest groups to make their life on the earth and to occupy vast habitats on soil and water.

They are found in wide range of climatic conditions, from -50^0 C to 40^0 C.

i. *Psylopa petrolei* is found in crude petroleum well.

ii. *Ephydra* fly inhabits great Salt Lake.

iii. Phytophagous insects feed on all types of plants.

iv. Saprophagous insects feed on decomposing materials.

v. Many carnivorous insects are either predator or parasitic on other animals or insects.

7. Size

Most of the insects are small conferring physiological and ecological advantages such as:

i. Exploitation of numerous ecological niches inaccessible to many other animals.

ii. For sustaining life and development, less space, food, time and energy requirements.

iii. Small insects can utilize maximum energy.

iv. They have less gravitational effect.

v. Due to tracheal respiration, muscular action is more effective.

vi. Easy escape from enemies.

8. Exoskeleton

The exoskeleton or outer covering is made up of a cuticular protein called Chitin. It is light weight and provides strength, rigidity and flexibility to the insect body. Chitin acts as:

i. External armour.

ii. Space for muscle attachment.

iii. Prevents water loss.

9. Resistance to desiccation

Insects body minimises the water loss due to following processes.

Prevention of water loss

i. Lipids and polyphenols present in the Epicuticle acts as water proofing.

ii. Wax layer with closely packed wax molecules preventing escape of water.

iii. Presence of closed spiracles prevent water loss.

iv. During egg stage, shell development prevents water loss and desiccation of inner embryo.

Conservation of water

i. Water is utilized during metabolism.

ii. Rectal re-absorption of water from faeces.

iii. Terrestrial insects use less quantity of water to remove the nitrogenous waste which is water insoluble (uric acid).

10. Respiration

i. Tracheal system ensures direct transfer of adequate oxygen to active tissues.

ii. Spiracles have closing mechanism and therefore, the air enters but water loss is restricted.

11. Reproduction

Reproductive potential of insects is very high. Following factors are responsible for their high reproduction:

i. Fecundity (egg laying capacity) is very high e.g. Queen termite lays 6000-7000 eggs per day for many years.

ii. Development period is short e.g. in corn aphid 16 nymphs are produced per female which reach adulthood within 16 days, this favours greater genetic changes in the insect population, like quick development of insecticide resistance.

iii. Selection of egg laying sites for protection of eggs.

iv. Parental care is exhibited (bees) and mass provisioning in wasps.

v. Presence of special types of reproduction other than oviparity and viviparity such as:

a) **Polyembryony:** Development of many individuals from a single egg e.g. parasitic wasps.

b) **Parthenogenesis:** Reproduction without male or without fertilization, e.g. aphids.

c) **Paedogenesis:** Reproduction by immature stages for example in certain flies.

12. Complete metamorphosis

More than 82 per cent of insects undergo complete metamorphosis (Holometabolous insects) with four stages:

i. **Egg:** Inactive stage; embryo develops inside.

ii. **Larva:** Active, feeds and digests, grows into many instars and stores food before pupation.

iii. **Pupa:** Inactive, internal reorganisation and resist adverse conditions.

iv. **Adult:** Active, reproduce and disperse.

The larval and adult foods are different so competition for food is less.

13. Defence mechanism

By using the defence mechanism, the insects escape from their enemies and survive. They may show the following features for defence:

i. **Behavioural:** Thenetosis- in which insect pretends to be dead; e.g. some beetles.

ii. **Structural:** In beetles hard forewing or elytra protects them from predation by birds.

iii. **Colouration:** Presence of protective colours which match with the environment; e.g. Stick insects.

iv. **Chemical:** Presence of defensive chemicals; e.g. Bees producing venom.

14. Locomotion

Insects use 3 legs and 2 wings, which give them different modes of locomotion, from one place to another, for their survival.

5

Zoological Nomenclature

Nomenclature should not be confused with terminology which deals with terms, like antenna, abdomen, radius etc. Names of animal and plants may be either vernacular or scientific.

Vernacular names or local names of plants and animals are different from region to region, from language to language.

Scientific names or Latin or International names are used universally all over the world. Without them we would not be able to communicate and to use the knowledge accumulated about animals and plants in literature. These international names are under certain rules or nomenclature.

Nomenclature

No-men-cla-ture means a system of names. It is derived from Latin *Nomen* means name and *calor* means to call, which means to call by names. Nomenclature is the "language of Zoology" and the rules of nomenclature are its grammar. Since all zoologists work with animals and used their names, it is essential that the general principles of zoological nomenclature be familiar to all zoologists.

"It is the application of distinctive names to organisms according to certain rules." The purpose of nomenclature is that a language must be widespread and the same words must have the same meaning to everyone. Therefore, universality and stability are the principal objects of any nomenclature. The role of nomenclature is to provide names for taxonomic categories in order to facilitate communication among biologists.

Scientific Nomenclature

The scientific naming of animals follows certain definite rules which are outlined in the "International Rules of Zoological Nomenclatures." and with the opinions handed- down by the "International Commission on Zoological nomenclature". These rules are concerned principally with the naming of groups from the family down. The names applied to systematic categories are usually derived from Latin or Greek and often refer to same characteristics of the animal or group named.

Zoological Nomenclature

January 1, 1758 is the starting date of all valid zoological names. The role of nomenclature is to provide names for taxonomic categories in order to facilitate communication among biologists.

- They are bound by law of availability.
- All names should satisfy provisions of code in order to be available for applying to taxon.
- They should be published as per law of priority.
- Earliest name published before 1.1.1758 should be applied but does not apply above family and below species.

The History of Zoological Nomenclature

Linnaeus for the first time secured the stability of names. This state lasted for almost 50 years. Linnaeus's follower enjoyed a similar authority. The main rule then was the two-word name and the principle of priority: The older name prevailed over junior one. However, even Linnaeus changed himself some names, some of which he regarded as more appropriate. With the Napoleonic and other wars, around 1800, the world became so disrupted that the scientists could not communicate and soon found themselves in a situation, when they travelled, that different names in different countries were used for the same species. To keep international status of the scientific names, it became necessary to propose a set of rules which would unify and stabilize a universal usage of scientific names.

The brief history of nomenclature is as follows:

- **Linnaeus (1751)** had already formulated personal set of rules.
- **Fabricius (1778)** followed with another personal code for entomological nomenclature, and **Rudolph (1801)** and likewise.
- **Strickland (1842):** Due to the disturbances caused by the Napoleonic war, there had been a drastic reduction, in exchange of scientific publications and periodicals. Therefore, countries were ignored and resulted in development of many local scientific nomenclatures. Later, due to this problem, British Association for the Advancement of science, appointed a committee to draw up a general set of rules for zoological nomenclature, the code of which is called "Strickland code". This formed the basis of all future codes.
- **W.H. Doll (1872),** thirty years later, was appointed by "American Association for the Advancement of Science", to obtain views from

working naturalists of America, regarding nomenclature. This is called Doll Code. This was not adopted by the Association.

- **1882:** Douville Code, based on Stickland Code, was adopted internationally by geologists and the American Ornithologists Union.
- **1889:** The first International Congress of Zoology in Paris adopted, in part, rules drawn up by Professor Raphael Blanchard.
- **1892:** Blanchard code discussed at the Second International Congress of Zoology in Moscow.
- **1895:** International Commission of five zoologists appointed at the third Congress in Leyden. Further members added later.
- **1901:** A report by the Commission was adopted by the 5th Congress in Berlin.
- **1905:** The report was published as a code in French, German and English. It was entitled *Regles internationales de la Nomenclature zoologique*. It contained no directives concerning types.
- **1907:** Amendments to Code at Boston Congress including provisions for generic types were reported.
- **1913:** Amendments were made to Code at Monaco Congress including recommendations regarding type of species.
- **1929:** Amendments to Code at Budapest Congress were made.
- **1930:** Code at Padua Congress amended.
- **1948:** Formal rules and recommendations regarding type specimens adopted at the Paris Congress.
- **1961:** Publication of 1st edition of the **International Code of Zoological Nomenclature**.
- **1964:** 2nd edition of the **International Code of Zoological Nomenclature**.
- **1972:** At the 17th Congress of Zoology in Monaco, responsibility for future Codes was transferred from the International Zoological Congresses to the International Union of Biological Sciences.
- **1985:** Publication of 3rd edition of the **International Code of Zoological Nomenclature**.
- **1999:** Publication of 4th edition of the **International Code of Zoological Nomenclature.**

It is evident that zoological nomenclature was an international matter and an international set of rules should be formed.

Rules of Nomenclature

After various attempts, in Britain, France, Germany, and especially after discussions at international Congresses, a set of rules was accepted. Later a special commission was setup at the international congresses of zoology and now a special International Commission on the Zoological Nomenclature exists. The present code has many predecessors, but it was mainly inforced since the 15th International Commission Congress of Zoology in London, 1958. Now a new edition of the Code is prepared.

The Zoological Code recognises three main types of names:

1. Family group names (FAM).
2. Genus group names (GEN).
3. Species group names (SP).

The other ranks are not controlled by the code.

- The names of genera, subgenera, species and subspecies are always Latinized; generic and sub generic names are nouns in the nominative or genitive and agree in gender with the generic name.
- A species or subspecies is usually referred to by a scientific name. The scientific name of a species consists of the genus and one trivial name and that of subspecies consists of the genus and two trivial names. Thus the scientific name of a species is a binominal and that of subspecies is a trinomial.
- Scientific names are always printed in italics (if written or type written, italics are indicated by underlining). Scientific names are followed by the name of the author (The person who described the species or subspecies); authours' names are not italicized. The genus name always begins with a capital letter, trivial names do not.
- If the authors name is in parenthesis, it means that he described the species (or subspecies in case of a trinomial) in some genus other than the one in which it is now placed. For example:
    i. *Papilio ajax* Linnaeus - The black swallow tail. The species *ajax* was described by Linnaeus who described it on genus *Papilio*.
    ii. *Diabrotica duodecimpanctata* (Fabricius) - The spotted cucumber beetle. The species *duodecimpunctata* was described by Fabricius, who described it in some genus other than *Diabrotica*, and this species has been transferred to the genus *Diabrotica*.
- A species referred to but not named is often designated simply by 'sp', for example *Gomphus* sp., refers to a species of *Gomphus*. More than one

species may be designated by 'spp.', for example *Gomphus* spp. refer to two or more species of *Gomphus*.

- The names of some systematic categories have standard endings and hence can always be recognized as referring to a particular sort of group. These are as follows:

i. Super Family names end in-oidea eg- Papilionoidea- butterflies.

ii. Family names end in idae – for example – Papillionidae – swallowtails and Parnassians.

iii. Sub family names end in – inae for example Papilioninae- swallowtails.

iv. Tribe names end in – ini for example Carbonini- square-headed wasps.

v. Other category names may have any sort of ending, at least in the case of groups.

vi. Family and sub family names are formed by adding – idae or –inae respectively, to the root of the name of the type genus (for example, the type genus of the swallowtail family is *Papilio*, the family name is Papilionidae).

vii. If a species is divided into subspecies, that particular subspecies, which includes the type of the species, has the same sub species name as species name (eg.- *Tetragoneuria cynosura cynosura*).

viii. Similarly, if a genus is divided into subgenera, the subgenus which contains the type species of the genus has the same subgenus name as genus name.

Origin of the Binomial System

Linnaeus published the 10th edition of the *Systema Naturae* in which he consistently used his system of binomial nomenclature. This system was adopted by other naturalists but with no rules governing it. By the middle of the 19th century there was confusion which increased by proliferation of new species resulting from the growth of Science and voyages of exploration.

The International Rules of Zoological Nomenclature

The International Rules of Zoological nomenclature or International code was adopted by the fifth International Zoological Congress (Berlin 1901). It consists of 41 articles and 20 recommendations, dealing with family, generic, specific and sub specific names with their validity, formation and orthography. Articles 33 to 41 dealt with priority, others with the designation of types and rejection of names. Various provisions of the International rules are explained in detail in chaps 12 to 16.

The adoption of the International Rules has helped not only to produce stability in nomenclature, but also to standardize certain taxonomic procedures. However, some changes have involved the revision of articles or the adoption of new articles.

More basic changes in the articles have been adopted by vote of the International Commission and after approval by the section of Nomenclature, by formal vote of the International congress in plenary session.

On four occasions major changes in the articles have been adopted since 1901. Three of which are given below:

I. The first was the refinement of the type method adopted by the Seventh International Zoological Congress at Boston. The principle involves, according to which the names of all categories up to the family are based on a type.

II. The second major change was the plenary powers Resolution (Monaco, 1913). It permits the suspension of the rules in any case where "the strict application of the rules will clearly result in greater confusion than uniformity".

III. The third major change was the modification of art 25 as adopted by the Budapest congress 1927. The original version of the rules failed to require a monitoring of the differentiating characters of the genus, species or subspecies in the formal description and in the case of generic name, the definite unambiguous designation of a type species. It was decided that it would be mandatory after Dec. 31, 1930 to include in the formal description a summary of characters which differentiate or distinguish the genus or the species from other genera or species and in case of genus, the definite unambiguous designation of a type species.

Functions and Powers of the International Commission

The International Commission on Zoological Nomenclature derives its authority from the International Congresses of Zoology, reporting at each meeting of the congress. It has full power to proceed about its business in inter congress periods. Results are published in the Bulletin of Zoological Nomenclature and other official publications to be recommended to the congress amendments or additions to the rules.

The functions are:

i. To comprise zoological nomenclature according to the Rules.

ii. To compile the official lists of generic and trivial names in zoology.

iii. To use the plenary powers to set aside the International Rules when it creates greater confusion than uniformity.

International Code of Nomenclature

After1800, virtually when all authors had adopted the Linnaean system, a new source of confusion appeared. Many authors decided to change the existing names if they had not been correctly formed according to Greek or Latin grammar. Also Geographical names were often changed when they were found inaccurate. Therefore, the need for a nomenclatural procedure was recognized.

Why we do not have names which would last forever and would not need to be changed. This is not possible due to three reasons:

1. Nomenclatural (Legislative)

Nomenclatural changes result from disagreement with the Code, which became a fairly complicated set of rules, the main being principle of priority. It begins with the definitions such as publication, its date, authorship, synonymy and homonymy of names which are separated in certain groups like species, genus and family names. Various species terms keep on introducing. Not every name is available and not every available name is valid in its name group.

2. Taxonomic

The taxa names fall into three categories: (a) species - group, (b) genus- group (c) family- group. A taxon must be published with the obvious aim for scientific record. A name is made available by being published under certain conditions and accompanied by a description or at least conditions to recognise the taxon concerned. Before 1931 a generic name was available without its description. But after 1931 a generic name with a type species is a must. In the past even anonymously published names were available, but now they are not.

3. Synonymy and homonymy

This includes taxonomic categories such as species, genus and family.

Synonymy means that within a category different names refer to same taxon and out of them only one is valid. Synonymy is more often linked to changes of taxonomic nature, but there are some nomenclatural changes as well. In case of species name, it is easy to find out the older published name, but not so in generic names. Then, the first published name becomes valid after a search for the exact date of publication.

Homonymy means that a name of same spelling was used twice or more in the same category. In such cases the older one is valid irrespective of used by same or different authors. Latin became the language of binominal scientific names. Latin also changed over a period of time and some spelling varied. Thus, some spellings had to be accepted restricted to few examples mentioned in Code (Art. 58) in the species- group names only. Example, *caeruleus*, *coeruleus* and *ceruleus* are regarded as identical species name. But similar difference in the genus- group would not be regarded as identical. For example, *niger*, *nigra* and *nigrum* are homonyms in the same genus but only one of them is right, according to the gender of the genus, e.g. *Carabus niger*, *Perla nigra*. The code makes the agreement in gender of a specific and generic name compulsory (Art. 34b). It also provides many examples of proper use of Latin and Greek names, because scholars of those languages are rare.

Principles of Zoological Nomenclature

There are six principles stated in the Code of Zoological Nomenclature.

1. The Principle of Priority or Law of Priority

"The principle that the valid name of a taxon is the oldest name applied to it, provided that the name is not validated by any provision of the code or by any ruling by the commission."

January 1, 1758 is the starting date of all valid zoological names. Of all the rules of zoological nomenclature, the most difficult to formulate was the one determining which of two or more completing names should be chosen. It covers the period from Jan 1, 1758 to the present. Its basis is to be found in Art 25 of the Rules and as amended in 1948. Its essential provisions are that the valid name of a genus or species can be only that name under which it was first designated on the conditions:

I. That (prior to Jan 1, 1931) this name was published and accompanied by an indication or a definition or a description.

II. That the author has applied the principles of binominal nomenclature.

III. That no generic name nor specific trivial name published after Dec. 31, 1930, shall have only status of availability (hence also of validity) under the rules, unless and until it is published either:

a) With a statement in words indicating the characters of genus, species or subspecies concerned.

b) In the case of a name proposed as a substitute for a name which is invalid by reason of being a homonym, with a reference to the name which is thereby replaced.

c) In case of generic name or sub generic name, with a type species designated or, as the case may be, indicated in accordance with one or other of the rules prescribed for determining the type species of a genus or subgenus on the basis of original publication (Rules (a) to (d) in Art 30.).

d) That even if a name satisfies all the requirements specified above, that name is not a valid name if it is rejected under the law of homonymy.

2. Principle of Binomial Nomenclature

Linnaeus in his 10th edition of *Systema Naturae* (1758), used double name uniformly and constantly for all kinds of plants and animals. First they are grouped to have an idea of interrelationships of living organisms. All insects which possess well defined and constant character of form and structure which interbreed are grouped under a species. These species are associated in a higher category, the genus or genera.

Therefore, "The scientific name of a species, and not a taxon of any other rank, is a combination of two names, the first being the generic name, and the second the specific name. The family and genus group are uninominal."

Trinomial Nomenclature

Some Entomologists contrary to present nomenclatorial practice, have used trinomials for other things besides geographical races, which they have called "varieties". Many such varieties are merely individual variants, due to food or climate conditions, or seasonal or colour forms, which should not have been designated by a trinomial. In zoological nomenclature "a trinomial is used only for a geographic race and varieties of other sorts" such as subspecies names.

3. Principle of Name Bearing Type

Each nominal taxon has actually or potentially its name bearing type, that provides the objective standard of reference by which the application of the same is determined.

4. Principle of Coordination

The family group, or the genus group, or the species group, a name established for a taxon at any rank in the group, is deemed to be simultaneously established with the same author and date for taxa based on the same name bearing type at other rank in the group.

5. Principle of Homonym

If an available name is a junior homonym of another available name it must be rejected and replaced.

6. Principle of the First Reviser

Relative precedence of two or more names or nomenclatural acts published on same date, or of different original spellings of the same name, is determined by the first reviser.

Specific Name

"The binominal combination of a generic name and a specific trivial name which constitutes the scientific designation of a species" (International

Commission 1948); also used by many workers and in the original rules, in place of trivial name.

Vernacular Names

There are local or general purpose names or common names, that are not proposed for zoological nomenclature. An organism may have many names in local languages. But different nations need to have a single name to be able to use it for the same taxon, to avoid ambiguity. This is one of the reason that scientific names are required, governed by a code of rules.

Scientific Names

These are Latin or Latinized names for any organism. They can be derived from any language but must be converted to a Latin form using Latin letters (including the letters "j", "k", "w" and "y"). A species scientific name consists of two or three names, the genus name, the species name and the subspecies name.

Available Names

When a name is proposed and it meets the criteria of availability, it is added to the pool of names that are available for use. Available names are either valid or invatid.

- **Valid names (correct names in Bot. Code):** These are correct scientific names for taxa. They are not synonyms or homonyms. A valid name is not necessarily inviolate – which means that it is revised and synonymized, making it invalid by a taxonomist. A later taxonomist might unsynonymize the name hence making it valid again.
- **Invalid names:** These are incorrect scientific names for taxa. They can be subdivided as follows:

Subjectively Invalid Names

Names that have been made invalid as a result of a judgement that is a matter of opinion.

I. **Junior subjective synonyms (taxonomic synonyms in Bot. code):** Of a pair of subjective synonyms, a junior subjective synonym is the one that was described at the later date. A taxonomist might compare the types for two different names. He may make the more recent name a junior subjective synonym or the older name. This is a subjective decision because another taxonomist might have a different opinion and decide that the types are different species.

II. **Junior secondary homonym:** This is applied to species names only. In 1985, Bolton published a paper in which he (subjectively) synonymized *Triglyphothrix* with *Tetramorium* which had different characters. He then replaced the junior secondary homonym *Triglyphothrix silvestrii* by the new name *Tetramorium surrogatum.*

III. **Conditionally suppressed names:** A taxonomist can apply to The International Commission on Zoological nomenclature to have a senior subjective synonym suppressed, thus reversing the order of priority. This usually occurs when the junior synonym is commonly used while the senior synonym is ignored and not in use for a long time. Suppressed name can be used as a valid name.

Objectively Invalid Names

When a name has been made invalid due to factual reasons.

I. **Junior objective synonyms** (Nomenclatural synonyms in Bot. code). Two names that have the same type.

II. **Junior homonym in the family group and genus group.** Two names that have the same spelling but which refer to different taxa. For example, there is a fish called *Acantholepis* while an ant is also having genus *Acantholepis.* In this case, the ant genus is a junior homonym and has been replaced by name *Lepisiota.*

III. **Junior primary homonym in the species group.** Primary homonyms occur when two authors each describe a different taxon but use the same genus and species names. A junior primary homonym can not become a valid name by being transferred to another genus.

IV. **Totally suppressed names.** The Commission of Zoological Nomenclature decides that an available name is never to be used as a valid name despite the fact that it might be a senior synonym or homonym.

V. **Partially suppressed names.** Confusingly similar term to totally suppressed names.

Unavailable Names

Scientific names that do not fit the criteria of availability. An unavailable name is known as a *Nomen nudum.*

According to the latest Code (4th edition), an author should not displace a name that has been regularly used as valid (by at least 10 authors in 25 publications during the past 50 years covering at least 10 years) by an earlier synonym or homonym which has not been used valid since 1899.

6

Zoological Classification

Classification

Classification is ordering of organisms into groups (or sets) on the basis of their relationships that are of associations by contiguity, similarity or both.

Classification of living things is an attempt to interpret nature. It attempts to bring together the kinds that are alike and closely related and to separate those that are unlike and unrelated.

Animals might be classified in various ways but the classification is based primarily on structural characters. Those animals with certain structures in common are classified into one group, and those with other structures into other group. Thus the animal kingdom is divided into a dozen or so major groups called phyla (singular, phylum); each phylum has a name, and its members have certain structural characters in common. The characters used to distinguish phyla include the number of cells, symmetry, body form and segmentation, the nature of the appendages, and the arrangement of the internal organs.

On the basis of degree of complexity, and probably evolutionary sequence, the animal phyla are usually arranged in a series from the lower phyla to the higher ones. The principal phyla of the animal kingdom are as follows:

Phylum

1. Protozoa – Single celled.
2. Porifera – Sponges.
3. Coelenterata – Jelly fish, hydroids, corals, sea anemones.
4. Platyhelminthes – Flatworms, planarians, flukes, tapeworm.
5. Nematehelminthes – Round worms.
6. Trochelminthese – Rotifers.
7. Brachipoda – Brachiopods.
8. Bryozoa – Moss animals.
9. Mollusca – Molluscs, clams, snails, octopi.

10. Echinodermata – Starfish, sea urchins, crinoids, sea cucumbers.
11. Annelida – Earth worms, marine worms, leeches.
12. Onychophora – Onychophorans, Paripatus.
13. Arthopoda – Crayfish, millipedes, centipedes, spiders, insects.
14. Chordata – Fish, amphibian, reptiles, birds, mammals.

In classifying all animals, they are segregated into small and large groups on the basis of their similarity or dissimilarities. The largest group is known as Phylum followed by Class, Order, Family, Genus, species and subspecies.

BRIEF History of Classification

Hippocrates (460 – 377BC) and Democritus (460 – 370BC) observed animals.

Aristotle (384 – 322 BC) a Greek Scientist published book on History of animals in nine parts entitled "*Historia Animalium*" and described 500 animals according to their morphological characters, habits and habitat, that is why he is called as "Father of zoology". He has classified the animals on the basis of biological Taxonomy as under:

- Vertebrates: With red blood.
- Viviparous: Man, whale and other mammals.
- Oviparous: Birds, amphibians, reptiles etc.
- Invertebrates: Not red blooded.
- Cephalopods.
- Crustaceous.
- Insect, spiders etc.
- Mollusca and echinodermates.
- Sponges etc.

Pling (23 – 79) wrote a book in parts entitled "Natural history".

Galen (130 – 200) experimented on the physiology of internal organs of some animals. After this up to 1000 years no work was done on zoology. This was supposed as Dark Age of Zoological science.

Thiophrestus (370 – 285), a student of Plato and Aristotle described 480 plants in his book *Historia Plantarum* and is called "father of Botany".

Albertus Magnus (1193 – 1280) differentiated monocots and dicots in plants.

Otto Brunfels (1467 – 1554), a German first recognized the difference based on presence or absence of flowers and called them *perfecti* and *imperfecti*.

Jerome Bock (1498 – 1554), another German classified the plants into tree, shrubs and herbs.

Andrea Cesalpino (1519 – 1603) classified plants on the basis of fruits and seeds.

Gaspard Bauhin (1560 – 1624) was the first to find out the binomial classification in plants.

In 13th century **Albertan Magnus** again initiated work on animals and wrote a book on animals. During this century microscope was invented which helped a lot in the study of zoology and various aspects were studied by the following scientists:

Robert Hook (1635 – 1703) described the cells.

John Ray (1627 – 1705) was the first biologist who classified with the modern concept of species. He classified plants, (1704) insects, reptiles and animals (1693).

Francis Willoughby (1635 – 1672) died premature, was the first person who recognized the difference between genus and species.

Antony Van Leeuwenhoek (1632 – 1723) studied Protozoa, Bacteria, yeast etc. and is supposed to be father of microbiology.

Buffon (1707 – 1788), in his book "*Histoire Naturalle*" which has been published in 36 parts, has described about plants and animals in detail.

Carolus Linnaeus (1707 – 1778) is the first biologist who suggested "Binomial Nomenclature" in 1749 that is why he is called as "Father of Modern Taxonomy". In the book "*Systema Naturae*" which has been published in 1735, he classified plants and animals. He described the animal kingdom up to species on the basis of structural characters and gave species as a distinct name. 10th edition of this book was published in 1758 which marks the beginning of the consistent application of "Binomial System of Classification".

Georges Cuvier (1769 – 1832) referred as founding "Father of Palaeontology" was a French Zoologist who established the sciences of comparative anatomy and palaeontology.

Jean Baptiste deLamarck (1744 – 1829) gave his opinion on development of animals in his book "*Philosophie Zoologique* – 1809".

Charles Robert Darwin (1809 – 1882) studied the evolution of animals in 1866.

Gregor Jahann Mendel (1822 – 1884) formulated the principles of genetics in 1866.

Ernst Haeckel (1866) introduced the method of representing phylogeny by means of trees or branching diagrams. This was an exciting period in the history of taxonomy.

Kinds of Classification

The earliest record of Classification shows that arrangement of animals in biological classification has been described in various ways which may belong to any of the following types:

1. Phenetic classification.
2. Natural classification.
3. Phylogenetic classification.
4. Evolutionary classification.
5. Omnispective classification.

1. Phenetic Classification

The classification is based on observed characters without direct relation to phylogeny. The taxa are classified on the basis of few or all characters. If based on few characters, the subsequent change may occur, subject to discovery of natural affinities of the taxa. Acanson in the year 1757 first discussed the classification based on characters. With the advent of computers, the idea of numerical classification was extended by Sneath, Sokal and Moss. They have devised various methods of classification using similarity and dissimilarity of a large number of characters simultaneously.

This approach is quite useful for groups with immature classifications. The computer methods of phenetics can be a great help to taxonomy if combined with philosophy of evolution of evolutionary taxonomy together with the proper characters.

2. Natural Classification

Natural classification is a phylogenetic classification which reflects the evolutionary relationships among the groups that comprise it. Natural classification is one in which the groups are recognised by having a maximum number of attributes in common with their limits set by discontinuities in the diversity and capable of yielding the maximum number of correct deductions about correlations of other features.

3. Phylogenetic or Cladistic Classification

Phylogeny is an important part of classification and explains the associations involved in classification. Phylogenetic or cladistic classification is a type of systematics developed by Willi Hennig (1950) who discovered a more objective method of classifying organisms. Cladistic group of organisms are based on shared derived characters, not the overall similarity of group of members. Cladistic classification is exclusively based on phylogenetic branching. Cladistic phylogeny is opposite to numerical phenetics. It attempts to map the sequence of phyletic branching through a determination of shared primitive and shared derived characters.

Recently Christoffersen (1995) has described cladistics and phylogenetic classifications as two distinct classifications. In cladistics classification a cladogram is used as a graphical model for constructing biological system. A cladogram is a predominating bifurcating, asymmetrical nontruncate dendrogram with no defined vertical and horizontal axis.

Phylogenetic taxonomy, like Linnean taxonomy was modelled on a phylogenetic tree rather than a cladogram and like its predecessor, perpetuates the use of morphology as means of recognising clads.

4. Evolutionary Classification

The whole concept of this classification is based on Darwinism. Darwin's ideas influenced the workers when they started believing that the groups are created through evolution.

Some workers consider evolutionary and phylogenetic classification similar to each other since they both are based on the features derived from a common ancestor. Simpson (1961) and others prefer evolutionary classification because it commonly needs information which is still largely phylogenetic but practically impossible to include in a tree diagram. It is a combination of phylogenetic branching with the amount of evolutionary divergence existing between different taxa. It is based on the evolutionary relationships of organisms, not only phylogeny. This classification provides foundation of all comparative studies in biology through the degree of genetic similarity existing between organisms and the phylogenetic sequence of events in their history.

This classification is thus useful for the groups of organisms which are result of divergent evolution. All living animal species are related to one another by way of evolutionary descent. This type of relationship helps us to establish correct systematic groupings.

5. Omnispective Classification

Blackwelder (1967) put forward the extension of the concept of natural classification which is quite realistic and pragmatic. In this all the readily available features of an organism are included but only those relevant for classification are used for grouping and distinction. This practice is used by most of the taxonomists at present for classification of animals.

7

Modern Classification of Insects Insects of Agricultural Importance Characters of Orders and Family

The foundation of modern insect classification was laid by Brauer in 1885. According to Brauer insects were classified as Subclass: Apterygogenea (pterous) and Pterygogenea (winged) based on:

- Presence or absence of wings.
- Mouthparts.
- Metamorphosis.
- Number of Malpighian tubules.
- The nature of wings.
- Thoracic segments etc.

The earliest classification of insects was based on habits and habitats and on certain gross anatomical features that define the general aspects of different kinds. In classifying insects, not only anatomical characters but also facts pertaining to physiology, biology, distribution, ecology and sometime cytology of the species is taken into consideration.

Other factors such as difference in time of appearance, food preferences, method of hibernation or even reaction to certain insecticides are indication of fundamental distinctions between the identical forms. With the all available information, behaviour and other life characteristics, insects are correlated with anatomical features to develop natural classification.

In forming keys, it is necessary to depend on the use of physical characteristics of the insect. Sometimes larval stages of the insect are most important and distinctive, therefore key is prepared on the basis of larval characters and the name of the insect is based on the adult characteristics.

There is an exception to the above rules as in case of white flies Aleyrodidae, the instars are so tiny that they are not seen in the adult stage. Therefore, the pupal stage that shows good characters is used for their classification. In Lepidoptera and Diptera also, nearly 3% of the described species are known in larval stage.

Insect Classification

Over one and a half million (One million living and about 12000 species of fossil insects) have so far been described and named. The classification of such a vast number of insects is a difficult job.

The system adopted presently is given below:

Superclass Hexapoda

1. CLASS. Collembola.

   ORDERS. Arthropleona, Neelipleona, and Symphypleona.

2. CLASS. Protura.

   ORDER. Protura.

3. CLASS. Diplura.

   ORDER. Diplura.

4. CLASS. Insecta.

I. SUBCLASS. Apterygota.

   ORDERS. Microcoryphia and Zygentoma.

II. SUBCLASS. Pterygota.

A. INFRACLASS. Paleoptera.

   ORDERS. Ephemeroptera and Odonata.

B. INFRACLASS. Neoptera.

a. DIVISION. Polyneoptera (orthopteroid orders)

   ORDERS. Orthoptera, Grylloblattodea, Dermaptera, Plecoptera, Embioptera. Dictyoptera, Isoptera, Phasmida, Mantophasmatodea, and Zoraptera.

b. DIVISION. Paraneoptera (hemipteroid orders).

   ORDERS. Psocoptera, Phthiraptera, Hemiptera, and Thysanoptera.

c. DIVISION. Oligoneoptera (endopterygote orders).

   ORDERS. Mecoptera, Lepidoptera, Trichoptera, Diptera, Siphonaptera, Neuroptera, Megaloptera, Raphidioptera, Coleoptera, Strepsiptera, and Hymenoptera.

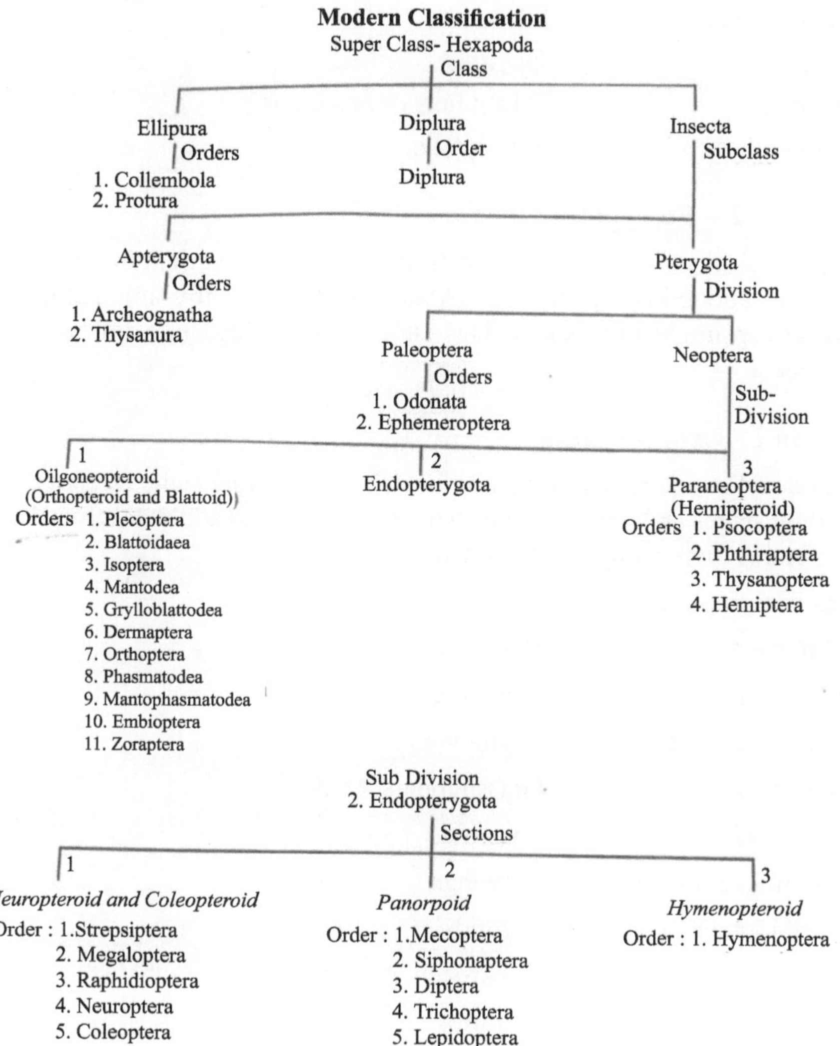

Insects of Agricultural Importance

Classification of insects is basically adopted by A.D. Imms (1957), according to him class Insecta or Hexapoda is divided into two sub classes:

Sub Class 1- Apterygota

Apterous insects, the wingless condition presumed to be primitive, metamorphosis slight or absent. Adult with one or more pair of pregenital abdominal appendages. Adult mandible usually articulating with the head capsule at a single point. This sub class is divided into following four orders:

1. Thysaneura - Silver fish
2. Diplura - Japygids
3. Protura - Telson tails or proturans
4. Collembola - Spring tails

Sub Class 2 – Pterygota

Winged or secondarily wingless insects, metamorphosis varied. Adults without pregentital abdominal appendages. Adult mandible usually articulating with the head capsule at two points. This subclass is divided into following two divisions:

Division I - Exopterygota

Wings develop externally. Metamorphosis simple, pupal stage rarely present. Immature stages are called nymphs which are similar to adults in structure and habit. In this division following orders are present:

5. Ephemeroptera - Mayflies
6. Odonata - Dragon flies
7. Plecoptera - Stone flies
8. Grylloblattoidea - Gryloblatta
9. Orthoptera - Grasshoppers, crickets
10. Phasmida - Phasmids
11. Dermaptera - Earwings
12. Embioptera - Embiids
13. Dictyoptera - Cockroach, mantids
14. Isoptera - Termites
15. Zoraptera - Zorapterans
16. Psocoptera - Booklice
17. Mallophaga - Birdlice
18. Siphunculata - Sucking lice
19. Hemiptera - Plantbugs, aphids, whiteflies, mealy bugs etc.
20. Thysanoptera - Thrips
21. Mantophasmatodea - Mantophasmatids (New order discovered in 2002)

Division II –Endopterygota

Wings develop internally meteramorphesis complex, pupal instar present, immature stages are called larvae which differ from adults in structure and habits. It is divided into following orders:

Order

22. Neuroptera - Lace wings, alderflies, snake flies etc.
23. Mecoptera - Scorpion flies
24. Lapidoptera - Butterflies and moths
25. Trichoptera - Caddish flies
26. Diptera - Two winged flies or true flies
27. Siphonoptera - Fleas
28. Hymenoptera - Ants, bees, wasp etc.
29. Coleoptera - Beetles and weevils
30. Strepsiptera - Stylops

Characters of Important Orders and Family with Examples

Order – Orthoptera (orthos = straight, pteron = wing)

Grasshoppers, crickets and locusts

Characters

1. Winged or brachypterous or apterous.
2. Mouth parts biting and chewing type (Mandibulate type).
3. Hind legs usually enlarged and modified for jumping.
4. Two pairs of wings, sometimes absent or vestigial, forewing straight, thickened called tegmina, hind pair of wing membranous.
5. Gradual metamorphosis, the nymphs resemble the adults in all essential features and habits.
6. A pair of unsegmented short cerci is present.

This order is divided into two sub orders:

Suborder- 1 – Ensifera

1. Antennae are longer than their body length and many segmented.
2. Tympanal organs (auditory organs) are located on the fore tibia of the leg.

Example – Long horned grass hoppers and crickets.

Suborder- 2 – Caelifera

1. Antennae are shorter than their body length with less than thirty segments.
2. The Tympanal organs are located at the sides of 1st abdominal segment.

Example – Short horned grass hoppers and locusts.

Family – Acrididae

1. These are moderately long insects with prominent head and legs.
2. Diurnal in habit.
3. The antennae is always much shorter than the body length.
4. The auditory organs are located on the sides of the Ist abdominal segment.
5. There is usually one generation in a year.

Example – Kharif grass hopper, *Hieroglyphus banian, H. nigroreplatus,* Desert locust, *Schistocerca gregarea,* Migratory locust, *Locusta migratoria.*

Order – Isoptera (Iso = equal, ptera = wing)

Termites or white ants

Characters

1. Moderate sized, thin skinned, social insects, consisting of several castes such as winged king and queen, wingless king and queen, workers and soldiers.
2. Metamorphosis simple.
3. Mouth parts of the typical biting and chewing type.
4. The wings are equal in size, long, narrow, membranous, some what opaque.
6. Workers and soldiers of both the sexes are wingless and sterile forms.

Family 1. Mastotermitidae eg.- *Mastotermes* spp.

Family 2. Kalotermitidae eg.- *Kalotermes* spp.

Family 3. Rhinotermitidae eg.- *Rhinotermes* spp.

Family 4. Hodotermitidae eg.- *Hodotermes* spp.

Family 5. Termitidae

Characters

1. Members are mostly subterranean and form a termitarium.

2. Wings only slightly reticulate, wing membrane and margin more or less hairy.
3. Pronotum of workers and soldiers narrow.
4. The queen attains enormous proportions, the increase effecting only the abdomen and not the head and thorax. This obesity is known as physagastory.

Example – Termite- *Odontotermes obesus, Microtermes obesi.*

Order – Hemiptera (Hemi = half, pteron = wing)

Characters

1. Two pair of wings usually present. The anterior pair most often of harder consistency than the posterior pair, either uniformly (Homoptera) or with the apical portion membranous than the remainder (Heteroptera).
2. Mouth parts piercing and sucking type.
3. Metamorphosis usually gradual.
4. The abdomen has no cerci.

Example – Plant bugs, leaf hoppers, coccids, white flies etc.

This order is divided into two suborders:

Suborder 1- Heteroptera (Heter = different, pteron = wing)

Characters

1. The fore wing thickened and lengthy basally and membranous apically known as hemelytra.
2. Metamorphosis is complete.
3. Body usually broad and flattened dorsoventrally.
4. A plate usually triangular in outline called scutellum, located between the bases of the wings.

Following are the important families:

Family 1. Pyrrhocoridae

i. Bugs are brightly coloured with red or black markings, elongate or oval in shape.

ii Hemelytra wings having membranous wing with five unbranched veins.

iii. Antennae four segmented, rostrum four segmented.

iv. Ocelli absent.

v. Coxa rotatory, tarsi three segmented.

Example – Red Cotton bug, *Dysdercus cingulatus.*

Family 2. Coreidae

i. These are dull coloured mostly brownish bugs.

ii. Hemelytra membranous with five unbranched veins.

iii. Body oval or elongate with narrow head.

iv. Hind femora and tibia have conspicuous enlargement or leaf like.

v. They produce pungent smell, therefore called gundhi bug.

Example – Rice gundhi bug, *Leptocorisa varicornis*

Family 3. Pentatomidae

i. Brightly coloured medium sized having pentagonal shape.

ii. Body covered with shield like wings.

iii. Antennae four or five segmented, base concealed by lateral margin of the head.

iv. Scutellum triangular and extends posteriorly and covers the wings.

v. Hemelytra well developed with 5-12 veins.

vi. They are phytophagous or predacious and produce disagreeable odour.

Example

a) Stink bug, *Aspongopus janus.*

b) Green bug, *Nezara virudula.*

c) Painted bug of mustard, *Bagrada cruciferarum.*

Suborder 2- Homoptera (Homo = similar, pteron = wing)

Characters

1. Two pairs of wings are usually similar in texture and each wing is practically of same thickness throughout.
2. Excretion of honey dew is common in many members of this suborder.
3. Metamorphosis usually gradual while in some cases complete metamorphosis is also found.
4. Mouth parts piercing and sucking type.

Following are the important families:

1. Cecadellidae (jassidae) (leaf hopper or jassids).

i. Slender with tapering body posteriorly. The body wedge shaped.

ii. Pronotum not prolonged backward.

iii. Hind tibia with two or more rows of spines.

iv. Ovipositor long sharp and needle like.

vi. Mostly vectors of plant diseases.

Example

a) Paddy leaf hoppers, *Nephotettix apicalis.*

b) Paddy leaf hoppers, *Nephotettix bipunctatus.*

c) Mango leaf hopper, *Idiocerus atkinsoni.*

d) Cotton jassid, *Amrasca beguttula.*

2. Lophopidae (Fulgoridae) (Pyrilla)

i. Face being longer than wide with at least two lateral ridges.

ii. Hind tibia with some spines.

iii. Lateral ocelli present below the compound eye and slightly in front of it.

iv. Wings broad and held somewhat flat, often patterned.

v. Nymph with two tails.

Example – Sugarcane Pyrilla, *Pyrilla perpusilla*

3. Aleyrodidae (white flies)

i. Very minute insects, wings and body covered by powdery dust.

ii. Antennae 7 segmented, long.

iii. Last abdominal segment having vasiform orifice.

iv. Tarsi terminating into pad like empodium or spine in between the claws, 2 segmented.

v. Three nymphal instars, first mobile while next two immobile.

vi. The pupal shell ruptures into 'T' shaped slit to release the adult.

Example

a) Sugarcane whitefly, *Aleurolobus barodensis.*

b) Cotton white fly, *Bemisia tabaci.*

c) Citrus white fly, *Dialeurodes citri.*

4. Aphididae (Aphids, green flies or plant lice)

i. Small soft bodied insects with pear shaped body.

ii. Antennae long, tapering distally.

iii. Tarsai three segmented with a pair of claws.

iv. A pair of cornicles on 5th or 6th abdominal segment, which secrete honeydew to which attract the ants.

v. Polymorphism is found in adult winged and wingless forms.

vi. They reproduce parthenogenetically.

Example

a) Mustard aphid, *Hydophis (Lipaphis) erysimi.*

b) Bean aphid, *Aphis craccivora.*

c) Cotton aphid, *Aphis gossypii.*

d) Potato aphid, *Myzus persiki.*

5. Coccidae – (Mealy bugs and scale insects)

i. Inconspicuous, minute insects with highly specialised qualities.

ii. Body covered with powdery or waxy coating.

iii. Males alate with first pair of wings well developed, second pair reduced to halteres while females apterous.

iv. Tarsi one segmented with a single claw.

v. Larvae degenerate type, scale like, with functional mouth parts.

vi. Reproduction bisexual or parthenogenetic.

Example – Mango mealy bug, *Drosicha stebbingi*

6. Lacciferidae (Lac insect)

i. The insects are sluggish and have a sedentary life.

ii. They live inside the chambers in the tree twigs.

iii. They are degenerated, mostly without wings but with distinct legs.

iv. Males may be winged or without wings, mouth parts absent so they do not take food.

v. Female is larger and has a bag like body and secretes bulk of lac used commercially.

vi. Female lacks eyes, wings and legs and never moves after settling on twig.

vii. Nymphs after settling, cover them with a resinous substance secreted by dermal glands present all over their body.

Example – Lac insect, *Laccifer lacca.*

Order – Lapidoptera (Lapido = scales, pteron = wing)

Moths, butterflies and skippers

Characters

1. Insect with two pairs of membranous wings, but not transparent, covered by minute overlapping scales.
2. Mouth parts greatly reduced possessing siphoning type of mouth parts.
3. Metamorphosis complete.
4. Larvae Known as caterpillar or semilooper which possess biting and chewing type of mouth parts.

This order is divided into two suborders:

Suborder 1-Heterocera – Moths

Characters

1. Mostly nocturnal in habit.
2. Antennae are of varied form, filiform, pectinate bipectinate etc.
3. The wings usually lie horizontally or roof like at the sides of abdomen.
4. Pupae very often protected by cocoon.

Following are the important families:

Family 1. Gelechidae (Galachiid moths)

i. Small moths of dull cryptic colours.
ii. Forewings narrower than hindwings.
iii. Apical angle of hind wing pointed.
iv. Larvae feed under concealment.
v. Labial palpi long and upcurved.

Example

1. Grain & flour moth, *Sitotroga cerealella.*
2. Potato tuber moth, *Phthoremoea operculella.*
3. Pink boll worm, *Pectinophora gossypiella.*

Family 2. Pyralidae (Pyraustidae) (Snout moth)

i. Small moths with labial palpi projecting forward into snout like structure.

ii. Forewings elongate, narrower or triangular.

iii. Vein Cu_2 absent in forewing and present in hindwing.

iv. Legs in adult slender and long.

v. Tympanum present at the base of the abdomen.

vi. Adult females provided with a tuft of hairs at the caudal extremity.

Example

1. Cotton leaf roller, *Sylepta derogata.*
2. Jowar stem borer, *Chilo partellus.*
3. Sugar cane top borer, *Tryporyza nivella.*
4. Rice stem borer, *Tryporyza incertulus.*
5. Rice case worm, *Nymphula depunctalis.*
6. Sugar cane root borer, *Emmalocera deprecella.*

Family 3. Arctiidae (Tiger moth -Hairy caterpillars)

i. Moths with broad and stout body.

ii. Wings brightly coloured with bands or spots.

iii. Hind wing with both anal veins.

iv. Larvae densely clothed with long hairs.

Example –

1. Bihar hairy caterpillar, *Spilosoma (Diacresia) obliqua.*
2. Red hairy caterpillar, *Amsacta moorii.*
3. Castor hairy caterpillar, *Pericalia ricini.*
4. Sun hemp hairy caterpillar, *Utetheisa pulchella.*
5. Hairy caterpillar, *Euproctis lunata.*

Family 4. Nolidae (Tuft moth)

i. Small, dull coloured moths.

ii. Tuft of raised scales on the forewing.

iii. Larvae have muted colours and tufts of short hairs.

iv. Construction of ridged boat shaped cocoon that bears a vertical exit slit at one end.

v. Elongation of the forewing retinaculum into a bar like or digitate condition.

vi. Possession of a postpiracular counter tympanal hood.

Example – spotted bollworm, *Earias vitella, E. insulana, E. crumataria*

Family 5. Noctuidae (Borers, army worms, cut worms)

i. Moths dull coloured and cryptic

ii. Mostly nocturnal, attracted to light.

iii. Proboscis present, labial palps long and well developed.

iv. Maxillary palps small.

v. Vein Cu_2 absent in both wings.

vi. Larvae with prolegs bearing crochets, in semilooper first two pairs of prolegs absent.

vii. Pupation takes place in soil.

Example

1. Gram cutworm, *Agrotis ypsilon.*
2. Gram cutworm, *A. flamatra.*
3. Gram pod borer, *Helicoverpa (Heliothis) armigera.*
4. Tobacco caterpillar, *Spodoptera (Prodenia) litura.*
5. Cabbage semilooper, *Plusia orichalcea.*
6. Army worm, *Mythimna seperata.*
7. Fruit sucking moth, *Ophideres (Othris) conjuncta, O. fulonica, O. materna.*
8. Fruit sucking moth, *Calpe emerginata.*
9. Fruit sucking moth, *Achoea janata.*

Family 6. Sphingidae (Hawk moths).

i. Large robust flying moths.

ii. Antennae hooked apically.

iii. Proboscis extremely long, franulum well developed.

iv. Larvae with longitudinal oblique stripes. They possess horn like processes on 8th abdominal segment.

v. Pupation inside cocoon.

Example – Death hawk moth, *Acherontia styx.*

Family 7. Bombycidae (Silk moth)

i. Mostly silk producing moths.

ii. Antennae pectinate in both sexes.

iii. Wing lack frenulum in *Bombyx* while in *Eupterote* it is present.

iv. They pupate in dense silken cocoon.

Example – Mulberry silk worm, *Bombyx mori, Eupterote.*

Sub Order 2 – Rhopalocera – Butterflies and skippers

Characters

1. Diurnal in habit.
2. Antennae club shaped or clavate.
3. Wings remain vertical above the body.
4. Pupae are naked and they are called chrysalis.

Family 1. Papilionidae (Swallow tails).

i. Mostly large, extremely colourful butterflies with wings iridescent black with shades of blue, red, yellow or green.

ii. Antennae clubbed, fore tibia with pads.

iii. Hind wing with a tail like prolongation which is marginal extension of vein $M_{3.}$

Example- Lemon butterfly, *Papelio demoleus.*

Family 2. Nymphalidae (Brush footed, emperor butterflies).

i. Brightly coloured butterflies.

ii. Antennae clubbed.

iii. Forelegs reduced or vestigial, folded on the thorax.

iv. Tarsi or forelegs unjointed in males and 4-5 segmented in females.

v. Tibia covered with long hairs hence known as brush footed butterflies.

Example

Emperor butterfly, *Nymphalis antiopa.*

Painted lady, *Vanessa cardui.*

Pest of castor, *Ergolis merione.*

Pests of flowering shrubs, *Junonia hierta, J. orithya, J. lemonias.*

Family 3. Hesperidae (Skippers)

i. Small, short, stout, sombre coloured butterflies.

ii. They have habit of erratic darting flight that's why known as skippers.

iii. Antennae clavate with curved tip and widely separated at the base.

iv. Hind tibia with two pairs of apical spurs.

v. Larvae with five pairs of legs bearing crochets.

vi. Pupa attached to leaf from anal end by a silken thread.

Example – Paddy/ rice skipper, *Pelopidas mathias, Parnara mathias.*

Family 4. Lycanidae (Blue coppers).

i. Small, brightly coloured butterflies, sometimes with metallic shine.

ii. Male's forelegs reduced and lack claws.

iii. Tail like delicate prolongations on hindwings.

iv. Larvae mostly flattened and gland on them produce secretions that attract ants.

Example – Pomegranate butterfly, *Deudorix isocrates/ Virachola isocrates.*

Order – Coleoptera (Coleos = sheath, Pteron = wing)

Beetles and weevils

Characters

1. Two pairs of wing, fore wing thickened (hard & sclerotized) called elytra, hind wing membranous and protected by fore wing.
2. Both larvae and adult have biting and chewing type of mouth parts.
3. Metamorphosis complete.
4. Larvae of beetles are commonly known as grub. Snout beetle (weevil) grubs are legless (apodous).

This order is divided into following two suborders

Suborder 1 – Adephaga

Characters

1. Beetles mostly predatory in habit, they feed on other, insects.
2. Antennae generally filiform.
3. Notopleural suture is present.

4. The I[st] visible abdominal sternum is divided by the hind coxae and the posterior margin of this sternum does not extend completely across the abdomen.

Following are the important families:

Family 1. Cicindelidae (Tiger beetle).

i. Large, active beetles, brightly coloured, predaceous in nature.

ii. Eyes prominent, mandibles large and acutely toothed.

iii. Long legs adapted for running rapidly.

iv. Larva has relatively larger head, prothorax and mandibles.

v. Size up-to 10 to 25mm.

Example – Tiger beetle, *Cicindella sexpunctata.*

Family 2. Carabidae (Ground beetle).

i. Oval, broad and somewhat flat with metallic colours. There may be variations of size.

ii. Live under stones, ground and bark.

iii. The elytra are firmly attached together and wings atrophied.

iv. Legs slender, adapted for running and digging.

v. Adult and larvae are carnivorous.

Examples-

1. Egyptian predator beetle, *Anthia sexguttata.*
2. Carabid beetle, *Chlaenius bioculatus.*
3. Carabid beetle, *Calosoma indica.*

Suborder 2 – Polyphaga

Characters

1. The I[st] visible abdominal sternum is not divided by the hind coxae and the posterior margin of this sternum extends completely across the abdomen.
2. Hind trochanters are small.
3. Notopleural suture is absent.

Following are the important families:

Family 1. Dermestidae (Carpet/ Khapara beetle)

i. Small size adults, round or oval, ranging from 2-3 mm.

ii. Body covered with scales or setae.

iii. Antennae clubbed and fit into deep grooves.

iv. Hind femora fit into recesses of the coxa.

v. Larvae scarabaeiform and covered with setae.

Example – Khapra beetle, *Trogoderma granarium.*

Family 2. Curculionidae (curculioni = weevils or snout beetles).

i. Head modified into long rostrum and on its tip mouthparts are found.

ii. Rostrum well developed in females. The females bore holes by piercing rostrum.

iii. Labrum absent, palps reduced, mandibles absent or reduced.

iv. Antennae is clubbed, geniculate or clavate.

v. Larvae apodous and polyphagous.

Examples

1. Rice weevil, *Sitophilus oryzae.*
2. Gujhia weevil, *Tanymecus indicus.*
3. Sweet potato weevil, *Cylas formicarius.*

Family 3. Bruchidae (Seed beetles).

i. Insect small and stout, body covered with small setae or scales and pale grey or brown in colour.

ii. Adults free living and larvae are internal feeders of legume seeds.

iii. Head hypognathous and with snout like proboscis.

iv. Prothorax prominent somewhat triangular, notum greatly narrowed anteriorly.

v. Antennae clavate, serrate or pectinate.

vi. Elytra striated, short and do not cover the tip of the abdomen (pygidium).

vii. Legs short, hind femur thickened and toothed beneath, tarsal formulae 5-5-5.

Example – Pulse beetle, *Callosobruchus chinensis, C. maculatus.*

Family 4. Chrysomelidae (Flea beetles)

i. Beetles oval and convex with bright colouration.

ii. Antennae widely separated at base, eyes not prominent and head not produced.

iii. Prothorax laterally margined.

iv. Elytra covers entire body, pygidium not exposed beyond elytra.

v. Third tarsal segment not bilobed and forecoxae transverse.

Example

1. Red pumpkin beetle, *Raphidopalpa foveicollis.*
2. Rice hispa, *Decladispa armigera.*
3. Singhara beetle, *Galerucella bermanica.*

Family 5. Tenebrionidae

i. Beetles small, red or brown, body elongate and cylindrical.

ii. Hood like pronotum, concealing the hypognathous head.

iii. Antennae clubbed.

iv. Elytra smooth or sculptured sloping posteriorly.

v. Coxae of forelegs large and contiguous.

Example – Rust red flour beetle, *Triboliuam castaneum.*

Family 6. Coccinellidae (Lady birds)

i. Adults small or medium sized, oval or round, convex often brightly coloured with red, black yellow or brown spots or lines.

ii. Head partly concealed by pronotum.

iii. Tarsi 4 segmented, bi-lobed, second segment concealing the third.

iv. Mostly predaceous in the larval or adult stage on aphids, coccids, mites etc.

v. Some are phytophagous and destructive to crops.

Sub family – Coccinellinae (Lady bird beetles).

Example –

1. Lady bird beetle, *Coccinella septempunctata.*
2. Lady bird beetle, *Chilomenis sexmaculata.*

3. Lady bird beetle, *Rodolia cardinalis.*
4. Mexican bean beetle, *Epilachna varivestis.*
5. Potato lady bird or Hadda beetle, *Henosepilachna vigintioctopunctata.*

Family 7. Melolonthidae (Scarabaeidae) (Cock chafers, June beetles)

i. Variously coloured medium to large beetles with black, brown, green, blue or metallic body.
ii. Male antenna elongate serrate.
iii. Larvae feed on roots of plants while adult feeds on tree leaves.
iv. Beetles commonly attracted to light.

Example – White grub, *Holotrichia consanguinea.*

Family 8. Cerambycidae (Long horn beetles)

i Big sized beetles, body long and cylindrical with attractive colours.
ii. Prothorax narrow or as wide as mesothorax, usually spined or tuberculate.
iii. Stridulatory organs on the hind margin of prothorax.
iv. Antennae very long about one third of the body length.
v. Tarsi 5 segmented, tibia with 2 spurs.
vi. Larvae or grubs apodous, elongate, cylindrical, whitish and bore into tree trunks.

Example – Mango stem borer, *Batocera rufomaculata.*

Family 9. Elateridae (Click beetles or wireworms).

i. Adults typically nocturnal and phytophagous.
ii. Tibia with 2 spurs, antennae 11 segmented, pectinate inserted near the eyes.
iii. A spine on the prosternum is snapped into a corresponding notch on the mesosternum, producing a 'click' sound.
iv. Click beetle larvae are called wireworms. They may be saprophagous or phytophagous.
v. Few species are bioluminescent (Adult and larvae).

Example – *Pyrophorus*, Wheat wireworm, *Agriotes mancus.*

Order – Hymenoptera (Hymen = membranous, pteron = wing)

Bees, wasp, sawflies and parasitic wasp.

Characters

1. Wings typically four, small and membranous, hind pair of wing smaller, wing venation highly specialized.
2. Mouth parts biting and chewing or chewing and lapping type.
3. Metamorphosis complete.
4. Abdomen of female usually provided with a saw or piercing organ or sting.
5. Larvae either caterpillar like or grub like or legless.

This order is divided into two suborders:

Suborder 1. Symphyta (Saw flies).

Characters

1. These insects are characterized by the abdomen being broadly joined to the thorax with no marked constriction between the Ist and IInd abdominal segments.
2. Ovipositor adapted for sawing or boring but never a sting.
3. larvae (grubs) are caterpillar like which possess well developed thoracic and abdominal legs.
4. Prolegs without crochets.

Family. Tenthredinidae (True Saw flies).

i. Small to medium sized brightly coloured wasps.
ii. Fore tibia with two apical spurs.
iii. Saw like ovipositor with serrated plates to lay egg.
iv. Larvae caterpillar like which feed on flowers and leaves of vegetables.
v. Larvae with 2-8 pairs of prolegs without crochets.
vi. Pupation takes place in an elongated silken cocoon or earthen cells.

Example – Mustard sawfly, *Athalia lugens proxima.*

Suborder – Apocrita (Bees, wasp, parasitic wasp etc.).

Characters

1. These insects are characterized by a deep constriction between the propodeum (Ist abdominal segment) and the IInd abdominal segment called petiole or waist.

2. The larvae (grubs) are apodous (legless) or grub like with head and mouth parts reduced.
3. They possess a well-developed ovipositor with sting.

Family 1. Apidae (Bees)

i. Social insects with caste system consisting of queen, drone and workers.
ii. Mouth parts chewing and lapping type, labrum broader than long.
iii. Forewings with three submarginal cells, the third oblique. Radial cell one.
iv. Hind pair of legs modified for pollen collection.
v. Hind tibia without spurs, fore and mid tibia with apical spines.
vi. Ovipositor well developed, modified as sting.

Example

1. Bush bees, *Apis florea.*
2. Rock bees, *Apis dorsata.*
3. Indian honey bee, *Apis cerana indica.*
4. Italian bee, *Apis mellifera.*

Family 2. Elasmidae

i. Small brown or black wasps with compressed hind legs.
ii. Mostly primary or hyper- parasitic on Lepidoptera and Hymenoptera.
iii. Hind coxae greatly enlarged and wavy lines or diamond shaped patterns of setae on the hind tibia.
iv. Often, scutellum with elongated, overhanging flange.

Example – *Elasmus brevicornis.*

Family 3. Trichogrammatidae

i. Very minute insects with three segmented tarsi.
ii. Parasitic on eggs of large number of Lepidopteran pests.
iii. Forewings broad with rows of microscopic hairs.
iv. Forewings typically stubby and paddle like with a long fringe of hinged setae around the outer margin, to increase the surface area during the down stroke.
iv. They are not strong fliers and move along the wind.

Example – *Trichogramma chilonis, T. minutum, T. japonicum, T. brassicae, T. evanescence.*

Family 4. Chalcididae (Chalcid wasp)

i. Minute, small metallic coloured insects.

ii. Primary or secondary parasites of eggs, larvae and pupae of insects.

iii. They are characterised by having hind femur stout and swollen with rows of short teeth.

iv. Wings not folded longitudinally at rest.

v. Ovipositor short and straight.

Example – *Brachymeria* sp.

Family 5. Braconidae (Braconid wasp)

i. Braconids are very small and stout bodied wasps.

ii. They are primary parasitoids, both internal and external, on other insects.

iii. Reduced wing venation, only one recurrent vein, cross vein 2m- Cu is absent in the forewing.

iv. Pupation within cocoon inside or outside body of host.

Example – *Bracon hebitor.*

Family 6. Ichneumonidae (Inchneumon wasps)

i. Small to medium sized insects which are parasidoid of larvae and pupae of Coleoptera, Lepidoptera and Hymenoptera.

ii. Antennae long, 16 or more segmented.

iii. Presence of trochantellus on hind femur.

iv. Pterostigma distinct, narrow costal cell on forewing.

v. Two distinct recurrent cross veins.

vi. Absence of anal lobe on hindwing,

vii. Ovipositor typically long and protruded from the body.

Example – *Isotima javanis.*

Order – Diptera (Di = two, pteron = wing)

True flies, houseflies, mosquitoes etc.

Characters

1. Insects with single pair of membranous wings, hind pair of wings modified into haltares.

2. Mouth parts piercing and sucking or sponging type.
3. Prothorax and metathorax small and fused with well-developed mesothorax.
4. Metamorphosis complete.
5. Larvae are apodous (legless) called maggots which have biting and chewing type of mouth parts.
6. Pupa are generally free or enclosed in a puparium.

This order has been divided into three suborders:

Suborder 1. Nematocera (Mosquitoes and Gall midge)

Characters

1. Antennae of image (adult) usually longer than the head and thorax, many segmented, majority of segments alike, not forming an arista or style.
2. Maggots (Larvae) are eucephalous type which have well developed head and horizontally biting mandibles.
3. Discal cell generally absent.
4. Maxillary palpi 4-5 segmented.

Family 1. Culicidae (Mosquitoes).

i. Small, slender, long insects without ocelli.
ii. Antennae plumose in males and pilose in females.
iii. Females having piercing and sucking mouthparts to suck blood while males feed on nectar.
iv. Wings fringed, scales along veins and posterior margin of wing.
v. Larvae aquatic with metanaustic arrangement of spiraculars.
vi. They are vectors of various human diseases.

Example –

1. *Culex fattigause.*
2. *Anophelese spp.*
3. *Aedes aegepti.*

Family 2. Cecidomyiidae (Gall midge).

i. Flies minute, delicate known to infest plants by making galls.
ii. Antennae long moniliform with conspicuous whorls.

iii. Wing venation reduced with few longitudinal veins, without cross veins.

iv. Tibia without spurs.

v. Last larval instar of most of the species possess a sternal spatula or breast bone on the ventral side of prothorax.

Example-

1. Rice gall midge, *Orseolia oryzae.*
2. Mango gall midge, *Dasyneura mangiferae.*
3. Linseed gall midge, *Dasyneura lini.*

Suborder 2 - Brachycera (Horse flies, robber flies etc.)

Characters

1. Antennae of imago are shorter than head and thorax, generally three segmented with the last elongate, arista present on terminal segment.
2. larvae are hemicephalous type which possess an incomplete usually retractile head and ventrally biting mandibles.
3. Discal cell almost always present.
4. Maxillary palpi one or two segmented.
5. Pupa is free.

Family 1. Tabanidae (Horse flies)

i. Adult fly with stout body and large head, extremely noisy during flight.

ii. Eyes golden green with purple marks; holoptic in males and dichoptic in females.

iii. Antennae three segmented, 3rd segment, annulate and arista absent.

iv. Females blood sucking on animals while males feed on nectar.

Example – Horse flies, *Tabanus maculicornis, T. speciosus.*

Family 2. Asilidae (Robber flies)

i. Long flies with bristles all over, proboscis horny and piercing.

ii. Forelegs modified for catching and piercing prey.

iii. A characteristic feature is tuft of hairs forming bearded face and protruding eyes.

iv. Pretarsus ending in bristle like empodium and large pulvilli.

v. Larvae cylindrical with small conical head; predaceous or scavengers.

vi. Pupa with girdles of spines dorsally on head and trunk.

Example – Robber flies, *Philonicus albiceps, Hyperechia* sp.

Family 3. Bombyllidae (Bee flies)

i. Adults large which mimic bees, usually quite hairy.

ii. Body robust and densely haired.

iii. Most species with long proboscis which can not be retracted.

iv. Head round with a convex face.

v. Male eyes are holoptic.

vi. Abdomen large.

vii. They hover near flower and feed on nectar.

iv. Larvae parasites or predators of other insects.

Example – Bee flies, *Bombylius major.*

Suborder 3 – Cyclorrhapha (Fruit flies, syrphid flies, vinegar flies)

Characters

1. Antennae three segmented with a dorsal bristle like arista.
2. Maxillary palpi one segmented.
3. The larvae are acephalous type, in which head is reduced and the mandibles are replaced into mouth hooks which are working in a vertical plane.
4. Pupa are enclosed in a puparium.

Family 1. Tephritidae (Fruit flies)

i. Small or medium sized flies, usually with spots or bands on wings.

ii. Subcosta bends apically forward at right angle and then fades out.

iii. Middle tarsus with spurs.

iv. Male with long flexible adeagus.

v. Female with well-developed horny ovipositor through which it inserts eggs inside the fruit or stem.

vi. Larvae feed on fruits or bore into stems or flower heads.

Example

1. Melon fruit flies, *Bactrocera cucurbitae.*
2. Mango fruit fly, *B. dorsata, B. diversa, B. zonata.*

Family 2. Syrphidae (Syrphid flies or Hover flies).

i. Adults moderate to large flies, often seen hovering or nectaring at flowers.

ii. Bee or wasp like, brightly coloured with yellow stripes or bands.

iii. Hover flies are distinguished from other flies by a spurious vein between radius and media.

iv. Larvae are predaceous on aphids and other Homopteran nymphs, aquatic maggots are called rat tailed as they have a very long breathing tube.

Example- Syrphid fly, *Episyrphus balteatus.*

Family 3. Agromyzidae (Leaf miner flies).

i. Small to minute blackish or yellowish flies.

ii. Vibrissae present, subcosta vestigial.

iii. Fore femur with ventral spine or bristle.

iv. Larvae are leaf miners, tunnel into stems, pods and roots.

v. They pupate into the larval mine or in soil.

vi. Larvae are characteristic in having sphaeroidal concentration of calcium carbonate in the Malpighian tubules.

Example

1. Pea leaf miner, *Phytomyza atricornis.*
2. Arhar pod fly, *Melanagromyza obtusa.*

8

History of Insect Systematics Phylogeny and Evolution, Zoogeographical Regions

Origin of systematics is as old as Vedas (1500 BC), Ramayana (1900 BC), Mahabharat (1400 BC) and Upanisad (350 BC) in which mention had been made about various animals.

History of Insect Classification

Linnaeus (1758-1766) in the 12th edition of *Systema Naturae*, divided insects into 9 orders-

1. Coleoptera (earwigs – Dermeptera).
2. Orthoptera.
3. Hemiptera.
4. Lepidoptera.
5. Neuroptera (ant lion, Sialis, termites and dragon flies).
6. Hymenoptera
7. Diptera.
8. Thysanoptera.
9. Aptera (wingless insects, spiders, scorpions, terrestrial crustacians and Myriapods).

This system not only brought together unrelated forms but also separated insects with common descent.

Fabricius (1775) improved this system by taking into consideration structure of mouth parts in forming different Orders. He recognised following orders:

1. Elentherata (Coleoptera).
2. Ulonata (Ephemera, Phryams, Apterygota, Pirlids, Neuroptera, Psocids, Panorpes, Hymenoptera, Termites and certain Crustacia).
3. Synistata (Orthoptera, Blattodea, Dermaptera).
4. Agonata (Scorpions and certain Crustacia).

5. Unogata (Libellula, Myriapoda and Arachnids).
6. Glossata (Lepidoptera).
7. Rhynchota (Hemiptera, Aphaniptera, Thysanoptera).
8. Antliata (Diptera, Anoplura, Mallophaga, part of Arachnida and Crustacia).

Clairville (1798) followed primarily the Linnaeus system of classification with slight modifications.

Lamarck (1744-1829) however was the first to clearly separate insects from Crustacia, Myriapoda and Arachnids.

Latreille (1831) recognised 12 Orders belonging to Aptera and Alata. These were:

1. Thysanura
2. Parasita
3. Siphonaptera (All Aptera)
4. Coleoptera
5. Dermeptera
6. Orthoptera
7. Hemiptera (All Allata- Elythroptera)
8. Neuroptera
9. Hymenoptera
10. Lepidoptera
11. Phiphiptera
12. Diptera (All Allata- Gynoptera)

The foundation of modern classification was laid by **Brauer** in 1885.

According to Brauer insects were classified based on • Presence or absence of wings • Mouthparts • Metamorphosis • Number of melpighian tubules • The nature of wings • Thoracic segments etc.

According to him, Class Insecta is divided into two subclasses:

1. Apterygogenea (apterous).
2. Pterygogenea (winged).

Sharp (1899) : Partly changed the Brauer's terminology and introduced.

1. Exopterygota - wings developed outside the body.
2. Endopterygota – wings remain internally until pupa.
3. Anapterygota – wingless.

Borner (1904) recognized the existence of ectognathous Thysanura and Diplura within Apterygota.

Berlese (1909) included Collembola and Protura under two subclass.

Shipley (1904) adopted Sharp's classification and provided a new system in which suffix "ptera extended to all orders"

Handlirsch (1908) included 4 classes:

1. Pterygogenea
2. Collembola
3. Compodeoidea (Diplura)
4. Thysanura

Recent Work on Classification

Extensive research into the fossil insects of Australia and North America caused the discovery of fossils which came in to existence for insect classification. Credit goes to evolve such system to **Handlirch** (1908 and 1926), **Tillyard** (1926) and **Martynov** (1925, 1938). **Tillyard** made minor corrections to **Brauer's** system recognizing 29 orders.

- During a period of more than 20 years, beginning in 1917, **Tillyard** (1918-1920) expounded his views on insect phylogeny, stemming from his work he made important contributions concerning the origin and relationships of many insect orders. Tillyard's work on the endopterygotes is particularly well known. In his work he showed that the Hymenoptera and Coleoptera (with the Strepsiptera) form two rather distinct orders, only distantly related to the other endopterygote groups which collectively formed the panorpoid complex. Within the complex, the Mecoptera, Trichoptera, Lepidoptera, Diptera, and Siphonaptera form a well-defined group, with the neuropteroid orders clearly distinct.
- **Hinton** (1958) made a strong case for excluding these orders entirely from the panorpoid complex and placing them closer to the Coleoptera.
- While Tillyard was concentrating on the phylogeny of the endopterygotes, his American contemporary, Crampton, was directing his efforts toward solution of the problems of exopterygote relationships, especially the position of the Zoraptera, Embioptera, Grylloblattidae, and Dermaptera. Following his anatomical study he discovered winged zorapteran *Zorotypus hubbardi.*
- **Crampton** (1920) concluded that the Zoraptera were related to the orthopteroid orders, and he placed them in a group (superorder Panisoptera) that also contained the Isoptera, Blattida, and Mantida. Later he revised his views and transferred the Zoraptera to the psocoid (hemipteroid) superorder, after consideration of their wing venation.

- In 1922 **Crampton** placed the Zoraptera in the order Psocoptera and suggested that it was from psocoid like ancestors that the modern hemipteroid orders evolved.
- Five years later he concluded that the grylloblattids were closer to the Orthoptera (*sensu stricto*) than the blattoid groups and made the Grylloblattodea a suborder of the Orthoptera. The modern view is that the grylloblattids are probably survivors of the protothopteran stock from which both the orthopteran and blattoid lines developed. Crampton considered that the closest relatives of the Embioptera were the Plecoptera, placing the two groups in the superorder Panplecoptera. In his early schemes Crampton also placed the Dermaptera in the Panplecoptera.
- Almost simultaneously in 1924 **Crampton** and the Russian paleoento-mologist **Martynov** (1925) proposed an apparently natural division of the winged insects on the basis of the ability to flex the wings horizontally over the body when at rest. In the Paleoptera (=Paleopterygota=Archipterygota) are the orders Ephemeroptera and Odonata whose members do not possess a wing-folding mechanism. It must be emphasized, however, that the two orders are only very distantly related through their paleodictyopteran ancestry. The remaining orders, whose members are able to fold their wings over the body, are placed in the Neoptera (= Neopterygota). The latter contains three natural subdivisions, the Polyneoptera (orthopteroid orders), Paraneoptera (hemipteroid orders), and Oligoneoptera (endopterygote orders).
- **Jaennel** (1938) recgnized 40 orders as against 29 orders; the additional 11orders include the eigth orders of fossil insects.
- **Essig** (1942) recognized 34 orders.
- This system was elaborated by **Brues, Melander and Carpenter (1954)** in their work on the classification of insects and divided insects into 34 orders.
- According to **Boudreaux** (1979) the primitive thysanuran *Tricholepidion gertschi* is considered to be distinct enough to warrant its own order.
- As suggested by **Hinton** (1981), however, members of the small Southern Hemisphere family Nannochoris- tidae are clearly set apart from the other scorpionflies, with which they have been traditionally grouped in the order Mecoptera, and further study may result in the family being placed in its own order (Nannomecoptera).
- **Kristensen** (1991) placed Diplura close to the Ellipura principally on the basis of the entognathous condition.

- **Kukalova-Peck** (1991), however, put more emphasis on features of the thorax, suggesting that they are true Insecta. Again, the monophyletic nature, or otherwise, of the Paleoptera is controversial.
- **Sharov** (1966) and **Kukalova-Peck** (1985, 1991) argued strongly that Ephemeroptera and Odonata had a common ancestor.
- **Kristensen** (1991), however, dumped the Odonata with the Neoptera, this assemblage thereby becoming the sister group of the Ephemeroptera. The status of the Polyneoptera likewise remains questionable.
- Some workers believe that this is a monophyletic group, while others insist that the group is polyphyletic, the term "polyneopterous" simply describing a grade of organization. Certainly the position of the Zoraptera is enigmatic, this small order having a mixture of orthopteroid and hemipteroid characters.
- One recent suggestion is that Zorapterans may be the sister group of the Embioptera, itself an order of uncertain affinity showing similarities with Plecoptera, Dermaptera, and Phasmida of all the major groups, the Paraneoptera is the one that is widely accepted to be monophyletic, though there is argument over whether the Psocoptera and Phthiraptera should be linked as a single order (Psocodea) or remain separate.
- Most modern authors also consider the endopterygote orders (except for the Strepsiptera) to be monophyletic, the two major sister groups being the Coleoptera-neuropteroids and the Hymenoptera- panorpoids.

Insect Phylogeny

Phylogeny, or phylogenetics, is the study of evolutionary relatedness among groups of organisms, which is discovered through molecular sequencing data and morphological data matrices. The term phylogenetics is derived from Greek word or term *Phyla* and *phylon*, denoting "tribe", "clan", "race" and the adjectival form "*genetikos*", is derived from word genesis means "origin", "source", "birth".

In fact, phylogenesis is the process, phylogeny is science of this process and phylogenetics is phylogeny based on analysis of sequences of biological macromolecules (DNA, RNA and proteins). The phylogenetics study is a hypothesis about the evolutionary history of taxonomic groups.

Insects are a highly diverse group of organisms with a worldwide distribution, thus having a complex evolutionary history. They have conquered every terrestrial environment and have many intricate interactions with a wide variety of organisms, including predator-prey relationships. The Taxonomy of insects is also included in phylogeny, i.e the classification, identification, and naming of species which helps in phylogenetics, but remains methodologically and logically distinct.

Evolution and phylogeny are inextricably intertwined. Both were born together in the earliest part of the 19th century. Before, it was generally accepted that organic life was the product of special creation. Evolution is change of life with time. The person who joined the link between evolution and phylogeny was the German biologist **Ernst Hackle**. **Handlirsch's** family tree of insects is a classic and set the stage for all later serious works in insect phylogeny. They assembled a remarkable collection of insect fossil ranging from older ones occurring in the Paleozoic to later ones approaching the present line.

Phylogeny of Insects Involves a Series of Stages

- The first is the appearance of primitive wingless insects, apterygotes of the **Silurian** period or earlier. Fossil evidences suggest divergent evolution of ectognathous and entognathous apterygotes.
- The second was the development of wings in the **Carboniferous** period. The pterygotes might have evolved from apterygotes through a long series of intermediate forms. The early or primitive winged insects had very simple wing articulation that allowed flight but the wings could not be flexed over the abdomen at rest. They were the dominant group of insects during the **Carboniferous** and **Permian** periods.
- The third major crucial step in the evolution of the insect wings was the development of wing flexing mechanism that enabled the insects to flex their wings posteriowards over the abdomen. This condition is known as **neopteran** condition. It facilitated the insects to run and hide from predators and move to niches. This condition is superior to **palaeopterous** winged insects in which wings are continually outstretched.
- Neopterous insects radiated rapidly and attained the dominant position among the class Insecta.
- The last major step was the development of complete metamorphosis during the **Upper Carboniferous** period which enabled the insects to benefit from the favourable aspect of formation of different habitats.
- During the **Lower Carboniferous** period, such insects might have undergone profound adaptive radiations such as evolution of tracheal system, a relatively impermeable cuticle and the fat body.
- The diversification might have enabled insects to interact with the contemporary diversification of angiosperm plants.

- The evolution of spiracular mechanism in relation to the development of flight might have counteracted the drying tendency of the ambient air.
- The fat body, particularly with regard to the storage of fat, enabled the insects to withstand long periods of desiccation.
- The size of the insects also played important role. Insects are large enough to overcome surface tension forces and at the same time small enough so that they fall from any height without any injury.
- Insects evolved in the **Permian** era or the period before which the fossils are very few. Nevertheless, from the forms it is found that apterygote insects are more primitive than the pterygota.
- Pterygotes arose in **Carboniferous.** The oldest pterygote insect known is Paleodictyoptera, showing many affinities with dictyoptera.
- Due to the presence of chitin, preservation is quite good. Most of the theories regarding this, are based on the studies of the morphology and embryology rather than the fossil records. Many theories have been put forward to explain phylogeny. According to them the ancestors of insects were Trilobites, Arachnida, Crustacea, Myriapoda.

1. From Trilobita

This theory was given by **Handlirsch** (1908). The trilobites were more primitive among arthropods and most of the groups are supposed to be originated from them. According to him, most primitive insect arose from palaeodactyoptera by the paranotal extension of thorax and of segments in abdomen. He thought that these processes are extension of tergal notum. Such extensions were present on all segments of trilobites. Besides, the number of postcepahalic segments vary and in some trilobites their number could be the same as in the insects. The Trilobite *Neoletus erratus* also have compound eye, ocelli, antenna and multi jointed cerci at the end of the abdomen. Handlirsch imagined that the fossil of palaeodictyoptera could fill the gap between trilobites and insects but the view is not accepted because:

I. There is great structural difference between trilobites and insects which could not be explained except on imaginary basis.

II. Tracheation and body fold of insects and trilobites differ so it has been shown that insects who were present before the carboniferous era simultaneously go with the trilobites and not originated with palaeodictyoptera.

2. From Crustaceans

The theory was put forward by **Hansen** (1911). According to this theory the Class Crustacea resembles to the insects in following points:

I. Antennae of both are homologous.

II. Second antenna of the crustacean is suppressed in the insect.

III. Mandibles of the crustacean articulate with the head by 2 processes. Similar processes are present in Thysanura and Machalids.

IV. First maxilla of crustacean resembles to those of the maxilla of insects.

V. Second maxilla is homologous to the labium of insects.

3. From Myriapoda

I. The occurrence of eversible sacs at the bases of appendages in the symphylan, some Diplura and lepismid genera.

II. The similarity in structure and development of heart and aorta in Myriapoda and Insecta and of the segmented vessels in Diplopoda, Chilopoda and Arthopteroid insects.

III. The similarity in the Malpighian and tracheal tubes.

IV. The occurrence of ecdysis by a split at the hind border of head in Myriapoda, Protura and Collembola.

4. From Symphyla

Although there is a general similarity between the Myriapoda and insects but not all groups of Myriapoda show equal affinity. Symphyla show the maximum number of chatacters in which they stand closest to insects, due to which most modern authors connect insects with this group. The following characters may be taken into account in this connection:

I. Occurrence of Y shaped epicranial suture in many symphylan and insects.

II. The similarity in the structure of post mandibulan appendages; maxillae and labium with the same chitinous units.

III. The 3 lobed hypopharynx consisting of median lingua and a pair of anterolateral super lingua occurs in symphylan and primitive insect like Diplura, Collembola and larvae of Ephemeroptera.

IV. According to **Snodgrass** (1957) the head apodemes of Symphyla are suggestive of anterior arm of the tentorium of Thysonura and Pterygota, since they give attachment to the same muscles.

V. Terminal cerci of Symphyla corresponds to those of insects and cerci of some symphylan like *Scolopendra* and Diplura like *Anajapyx* have similar spinning glands.

VI. Occurrence of stylets and eversible sacs on abdomen of Smphyla and Diplura and their anal glands open on the cerci.

VII. A premendibular segmented organ occurring in late embryo of Symphyla or the adults of some of its species is equivalent to a similar organ in the embryo of Orthoptera and also of Placoptera, Mallopaga, Coleoptera and Lepidoptera in which it is called sub oesophageal body.

VIII. **Tiegs** (1941-45) have shown that the Symphyla like insects have embryologically 14 segments.

IX. **Manton** (1953) has shown that visceral gland is also seen in Symphyla and not in the adult of Myriapodan group.

X. According to **Imms** (1965) Symphyla and *Anajapyx* also agree in the structure of their mouthparts, antennal musculature and legs.

In spite of these important similarities between Symphyla and insects, there are certain weaknesses in the theory of Symphyla in the origin of insects. Snodgrass feels that importance of advocating Symphyla as source of origin is little due to the differences in mandibular structures. Another important objection is the situation of genital openings, which in insects is near the posterior end of abdomen but in Symphyla, it is situated anteriorly. About this Tiegs (1940, 1945) and Manton (1956) are of the view that progoneate condition of Symphyla is a secondary specialisation and probably insectan stem arose from Symphyla before this secondary specialisation took place.

In view of the clear relationship between the two groups, however inferences can be drawn:

a) That insects and myriapod arose from a common stalk.
b) That the insects arose by the neotony from the myriapod like arthropods possibly also from Symphyla.

Regarding the evolution, **Tillyard** suggested the name Protoptera for a common ancestor in Silurian Period from which two lines; Proginiata and Opisthogeniata arose. From Proginiata developed Myriapoda and Diplopoda, whereas from Opisthogeniata developed Chilopoda and Insecta.

Protoptera

1. Proginiata - Myriapoda and Diplopoda
2. Opisthoginiata - Chilopoda and Insecta

Willi (1960), on the other hand suggested that the phenomenon is much more common in insects and there are two main type of neotony:

1. Those in which rate of body development is constant but the development of reproductive organs is accelerated like appendages on abdominal segments.
2. The abdomen of Collembola has only 6 segments.

The Trilobitean, Crustacean and Collembolan theories of insect origin are out of date and unacceptable. Therefore, it can be concluded that the evolution of insect took place by neotony of Symphyla.

Evolution

If we look at the fossil animals, we find that they usually start with primitive form and eventually die out with extremely specialized forms. Doll's rule of irreversible evolution was established on this observation. As **Simpson** and other have pointed out, evolution do not move undeviating towards specialization. In fact, a character can be lost again in a phyletic line and a similar or equivalent character can be reacquired. Specialization and de-specialization often alternate in evolution. Also, each taxonomic character may evolve to a large extent independently of other characters. For this reason, it is often misleading to consider their number of differences between two categories as indicating degree of differences.

Too often several characters are partially or completely correlated. For instance, the arboreal mode of living in a group of mammals will inevitably lead to changes in the locomotor apparatus that may affect nearly every bone and muscle in the whole body. A change of feeding habit in birds may result eventually in structural modifications of the bill, the tongue, the plate, the jaw muscle, the stomach and perhaps other features. All these characters are a single adaptive complex and should not be treated and considered as series of independent characters. A shift into a new adaptive zone may lead to a comparatively rapid structural reorganization in order to acquire the needed specialization.

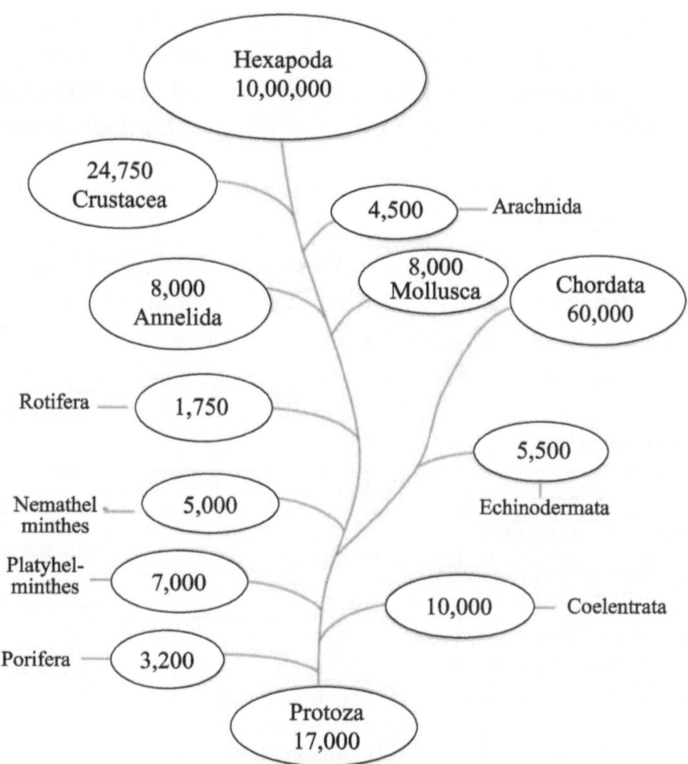

Phylogenetic tree of Animal Kingdom

(The area of the circles representing approximate numbers of species in the respective phyla)

Figure 8.1 Phylogenetic position of different phyla of animal kingdom

Evolution of Insects

- The evolution of insects dates back to the **Devonian** period, with the oldest definitive insect being the *Rhygniognatha hirsti*, estimated at 407 to 396 million years ago, possibly a flying insect.
- Global climate conditions changed several times during the history of the earth, along with it the diversity of insects.
- The Pterygotes underwent a major radiation in the **Carboniferous**.
- Endopterygota species underwent another major radiation in the **Permian.**
- Survivors of the mass extinction at the PT boundary evolved in the **Triassic** which are essentially the modern insect orders that persist to modern times.
- Most modern insect families appeared in the **Jurassic.**
- Further diversity in genera occurred in the **Cretaceous**.

- It is believed that by the **Tertiary**, there existed many of what are still modern genera; hence, most insects in amber are, indeed, members of extinct genera. It is believed that insects diversified in a brief 100 million years (give or take) into the modern forms that exist with minor change in modern times.

Zoogeographical Regions

It is believed that Insects originated in the **Devonian Period of Paleozoic era**. The geological History has been classified into following six eras which are further divided into periods. The periods of Cenozoic era are divided into epochs:

Table 8.1: Zoogeographical regions

S.No.	Eras	Periods	Epochs
1	Azoic (4500 million years ago)	-	-
2	Archeozoic (3500 million years ago) microfossils available (bacteria)	-	-
3	Proterozoic (1200 million years ago) fossils of algae, Protozoa, sponges	-	-
4	Paleozoic (600 million years ago)	(i) Cambrian (600) (ii) Ordovician (500) (iii) Silurian (425) (iv) Devonian (405) (v) Mississippian (345) (vi) Pennsylvanian (302) Or Carboniferous (vii) Permian (280)	
5	Mesozoic (230 million years ago)	(i) Triassic (230) (ii) Jurassic (180) (iii) Cretaceous (135)	
6	Cenozoic (70 million years ago)	(i) Tertiary (68)	(i) Palaeocene (70) (ii) Oligocene (35) (iii) Miocene (25) (iv) Pliocene (10)
		(i) Quaternary	(i) Pleistocene (2) (ii) Recent (0.011)

Note: The figures in paranthesis denote million years

1. The earliest known specimen in the history of insects is the **Devonian** fossil *Rhyniognatha hirsti*, dated between 396 and 407 million years ago. It was found in the Rhynei Chert formation, a well-preserved Devonian ecosystem which includes some of the first land plants with vascular tissues and among the earliest and best preserved fossils of terrestrial

fauna. The mandibles of this insect suggest that it had already developed flight, shrouding the origin of insect flight and other important aspects of the history of insects in mystery.

2. For tens of millions of years, insects and other small invertebrates were the only animals to colonize the land, at the time covered by short plants no taller than waist height. As plants grew and a lineage of fish evolved into the first amphibians, insects were joined by larger tetrapods, which would have consumed them in great numbers to survive. However, during the high oxygen levels of the **Carboniferous** period, about 320 million years ago, some insects grew to huge sizes, such as the griffin fly *Meganeura*, which had a two feet wingspan. But when oxygen levels went down, these insects promptly died due to an inability to circulate sufficient oxygen through their bodies.

3. The next major milestones in the history of insects occurred throughout the **Mesozoic**, when most modern groups, as we know them, evolved. Around 120 million years ago, flowering plants evolved, and the cooperation between insects (especially bees) and these plants led to a mutually beneficial evolutionary relationship. As a result, flowering plants are now the dominant terrestrial flora.

4. There are several known groups of Hexapoda (six-legged invertebrates) which are revolutionary basis to insects and would have split off from them prior to about 400 million years ago, when the first fossil insects appear. These include the abundant springtails as well as the less recognized proturans and diplurans. It is thought that springtails, proturans, and diplurans all evolved to form their hexapoda form of locomotion independently of one another, but only some insects gained the ability to fly.

9

Importance of Taxonomy

The science of taxonomy started with Linnaeus, who used two Latin names for an organism by a universally agreed name and identity, and further provided means for their realistic classification. These identifications and classifications got strengthened further with accumulated biological knowledge. In the recent times taxonomy has become more demanding subject for areas in entomology, which are dependent on access to accumulated experience and knowledge for their survival and development. By providing name and classifying an insect, taxonomy provides information on the nature of population, life cycle, habitat and behaviour and thereby developing hypotheses, concepts and theories, which are basic to biological control. In addition, it enables comparison and synthesis, all of which lead to refinement and fine tuning of the concepts and theories for identification.

Taxonomy contributes both directly and indirectly. The most important are as follows:

1. Theoretical Biology

Systematics has played a very important role in laying the foundation of some important fields of biology:

a) Population thinking: It is responsible for solving the problems of multiplication of species.

b) Understanding structure of species and of evolutionary role of peripheral populations.

c) Reaffirmation of role of natural selection as evolutionary factor in contrast to mutual theory of Mendel.

d) Understanding mimicry and other evolutionary areas.

e) Development of behavioural science.

f) Study of insect ecology to identify all the species of economic importance.

2. Applied Biology

Integrated pest management programme is the only way that our crops can be saved from insects and other pests. Such programme is designed to integrate chemical methods with the use of resistant plant varieties, predators and parasites, pheromones, hormones and lethal genes. All these methods are highly specific and can only be successful if the identity of the pests is accurately determined. Taxonomy has played a relying role in getting quick useful results.

3. Biological Control

The strategies undertaken are mutually dependent on science of taxonomy. There are three basic types of biological control:

a) **Conservation:** It involves preserving and enhancing natural enemies that are already present in the environment.

b) **Introduction:** It involves importing and releasing exotic (non- indigenous) natural enemies against foreign and indigenous pests.

c) **Augmentation:** Involves mass rearing of natural enemies in the laboratory and releasing them into the environment.

All these steps can be undertaken and successful only if taxonomy gets integrated with them, in handling the hosts as well as the natural enemies. Through taxonomy all these steps can be solved by the diagnostics of the pests that are causing crop losses and initiating and planning a long time strategy. By knowing the native home, its distribution and biological features, a taxonomist will answer all these questions by searching the already existing information about the pest, its bionomics, parasite, predators, vulnerable life stages, seasons of occurrence, proved biological control measures, economic injury level and economic threshold levels. Once the answers are available on the developmental stages of the pest that are vulnerable for applying management, and its natural enemies, we can forecast the strategy for biological control.

The next step in biological control is whether it would succeed. The answer again is taxonomy which helps in evaluating the pest incidence and parasitism. By identifying the insect and its life stages, its parasites and predators, these can be introduced and their efficiency can be evaluated.

Taxonomy can resolve the complex situation of host plant resistance, Bt transgenic based management measures which have led to development of biotypes, races and other intraspecific populations in many pests by orienting biological control strategies. The examples of worst affected pests due to host plant resistance are cotton boll worm *Helicoverpa armigera*, leaf feeders *Spodoptera litura*, white flies *Bemisia tabaci*, leaf hoppers *Nephotettix* spp., plant hopper *Nilaparvata lugens*, diamond back moth *Plutella xylostella* etc., which have developed biotypes and complex intraspecific populations. Thus, biological control of these problematic pests may go well in association with taxonomy to solve the difficult management strategy.

4. Agriculture and Forestry

Crops and trees are integral part of our biodiversity, but saving them from the harmful insects and pests is our utmost concern. The identification of their pests is necessary and for that taxonomy plays an important role. Every species harbour in its own niche in nature and differs from its related species in food preference, breeding season, tolerance to weather condition, resistance to parasites and predators, pathogens and competitors. If the identity of the pest is known, it is easy to obtain all such information from the existing literature.

When crop shows early signs of injury, samples need to be collected for all types of pests that are old or new to the farmer. There is always a possibility for a change of status of pest, which may be minor previously but has become major after some time. Some new pests may also be noticed.

The crop may show injury by leaf eating insects or sucking insects like plant bugs and aphids. Many other insects bore into plant stem or trees like wood boring beetles and their grubs, moth caterpillars etc. Some fruit flies cause damage in fruits and vegetables by laying eggs in the fruit rind causing enormous loss to fruit crops. This information is important before controlling any pest with the use of chemicals. On getting the current identity of the pest species, it becomes easier to collect information about its habitat which is vital for its effective control. The other thing which is taken into consideration, is whether it feeds on the upper or lower surface of the leaves. Similarly, many of the plant diseases are caused by certain vectors. The correct identification of a particular vector is a key factor for bringing the disease under control or kill its vector.

5. Wild life Management

Conservation of wild life and endangered species is also a big concern for which role of taxonomy is very important. Many programmes have been initiated all over the world to conserve fauna and flora. Many species of animals have become extinct and still many others are on the verge of extinction. Taxonomists help the environmental protection personals to identify all such animals including insects, which are endangered by man's activities. This is a challenging job in view of preservation and protection of our biodiversity.

6. Environmental Problems

Pollutants such as certain chemical pesticide residues may persist in the environment or even concentrate in certain plants or animals. The pollutants can be traced by identification of the species within the food chains. It is clear now that a biological approach requires a thorough understanding of the taxonomic relationships between pest species and component species of the ecosystem.

Water pollution is another major national problem. Each species has its own requirement for oxygen, nitrogen and other organic and inorganic compounds in water, certain algae, aquatic insects, helminth parasites of fish and some microscopic organisms are reliable indicators of the degree and nature of pollution.

Thus, the identification of species present in a particular location provides a quick and reliable monitoring system for detecting pollution.

7. Soil Fertility

Soil arthropods play an important role in increasing soil fertility. Many insects belonging to micro-arthropods such as Collembola, Protura and Diplura breakdown the litter into humus. Some of the Annelids tunnel in such a way that the earth becomes more aerated and enriched by their secretions and excreta. Such animals need their proper identification by the taxonomists.

8. Mineral Sequencing

The mineral sequencing of rocks and geological events in an area is useful to any search for fossil fuels and mineral deposits. The sedimentary rocks may be dated only by their enclosed fauna and flora. For this palaeontologists play a major role in the identification of fauna and flora and thus help in tracing the sequence of geological events. Its only through the fossils that their true sequence is determined as other method of faulting and folding of rocks is a complex process and pose a problem in correct identification.

9. Quarantine

Entry of new pests and diseases of plants, animals and human beings to many countries is a matter of serious concern. The spread of such organisms and disease carriers from one country to the other is through transportation of grains, plants, fruits, vegetables etc. Therefore, respective governments have established quarantine laboratories at aerodromes, ports etc. to check the spread of new pests and diseases. Taxonomists play a vital role in providing correct and prompt identification of the pest or disease. Unless this is done, no further action is possible in such matters.

10. Public health

A number of diseases are spread through arthropods, birds and mammals. Therefore, public health programmes are dependent on taxonomists to plan the eradication of the disease by targeting the concerned species. This is possible only by correct identification of the causal organism. For example, some species of mosquitoes are responsible for transmitting malaria and others are not. *Anopheles maculipennis* complex comprises many sibling species of which few are responsible for transmitting malaria. The control measures should be applied to the target species and therefore, malaria can be brought under control. Thus correct identification ensures a maximum effective control at a minimum cost.

11. National Defence

Information regarding disease vectors and parasites is an obvious application of systematics to national defence. The use of such biological means in the war are economical and require less efforts in operation. But they are dangerous to the human being, therefore, in such cases, both making the bomb and their destruction needs correct identification of the causing organism. Moreover, the identification of potential disease vectors is vital to the health of military and civilian populations all over the world.

12. Commercial Use

Some productive insects provide good source of commercial products such as honey, silk, lac and dyes that are integral part of our life. Besides, some other animals, either directly or indirectly provide food or eaten directly by men. Taxonomists can play an important role in improving the quality and quantity of the product by identifying the improved strains of the species. Information about the identification and biology of a species, can be a task of taxonomist to introduce it in other countries for commercial purpose. The examples of Italian honey bee *Apis mellifera* and European carp *Cyprinus carpio* are two well-known species that have been successfully established in India.

10

Numerical Taxonomy

Man has tendency to classify everything in a hierarchical manner for example, ships may be classified on size, weight, what they carry or naval versus merchant or passenger. Any of these classification is as valid as any other.

But classification of biological species is unusual in that a predefined system exists i.e. the evolutionary relationships among organisms. But evolutionary relationships can only be estimated; and contrary to popular belief, even a good fossil record does not conclusively show the phylogeny or evolutionary tree of a group species.

Numerical taxonomy can be defined as "Numerical evaluation of the affinity or similarity between taxonomic units and ordering of these units into taxa on the basis of their affinities". The term includes drawing of phylogenetic inferences from the data by statistical or other mathematical methods to the extent to which this should prove possible. The practice involves a number of fundamental assumptions and philosophical attitudes toward taxonomic work.

Numerical taxonomy is based on the basic idea put forward by Adanson. They may be called Adansonian and are described by following rules, modified by Sneath, 1958:

1. The ideal taxonomy is that in which the taxa have the greatest content of information and which is based on as many characters as possible.
2. Every character is of equal weight in creating natural taxa.
3. Overall similarity or affinity between any two entities is a function of the similarity of the many characters in which they are being compared.
4. Distinct taxa can be constructed because of diverse character correlations in the groups under study.
5. Taxonomy is strictly an empirical science.
6. Affinity is estimated independently of phylogenetic considerations.

When computers first came into existence for taxonomists, in the late 1950s, some biologists decided that opportunity existed to form methods for classifying objectively. Because, phylogeny could not be accurately discovered, it was decided to classify by 'resemblance' alone; even when convergent evolution could cause close resemblance. These techniques are known as PHENETIC methods.

1. A simple PHENETIC Method

The first step is to code each character into STATES e.g. presence of ocelli and absence of ocelli are each states of character ocelli. For analysis, these states may be given numerical numbers such as '0' for absent and '1' for present. For simplicity we can consider only two states (0/1) characters. The data for a set of 5 species (A- E) with 6 characters may be as follows: (Taxa at any other level than species can also be analysed by this method)

	1	2	3	4	5	6
A	1	1	1	0	1	1
B	1	1	1	0	1	1
C	0	0	1	0	1	0
D	0	0	0	1	1	0
E	0	0	0	1	1	0

The next step is to measure the resemblance between each pair of species. This can be expressed as a DISTANCE value between each species; distance (D) equals the sum of the differences in states of all the characters, for example:

Species **A** 1 1 1 0 1 1

Species **C** 0 0 1 0 1 0

Distance D = (1- 0) + (1 -0) + (1 -1) + (0 - 0) + (1 - 1) + (1 - 0) = 3

Some methods use SIMILARITY which is a count of the number of states which are the same in both species. In this case similarity also equals 3. However, distance is more versatile in that it can easily be applied to values other than 0 and 1, such as measurements.

A table of resemblance values is formed and this is analogous to a road atlas mileage chart, in that it tells us the distance between any pair of the species.

	A	**B**	**C**	**D**	**E**
A	-	0	3	5	5
B		-	3	5	5
C			-	2	2
D				-	0
E					-

A method will now be used to group or CLUSTER the species, which is called SINGLE LINKAGE CLUSTER ANALYSIS. It can be seen that species A and B are identical within the bounds of the sample of 6 characters and these species form the first cluster:

Species D and E form a similar cluster:

Species C has a distance of 2 to D:

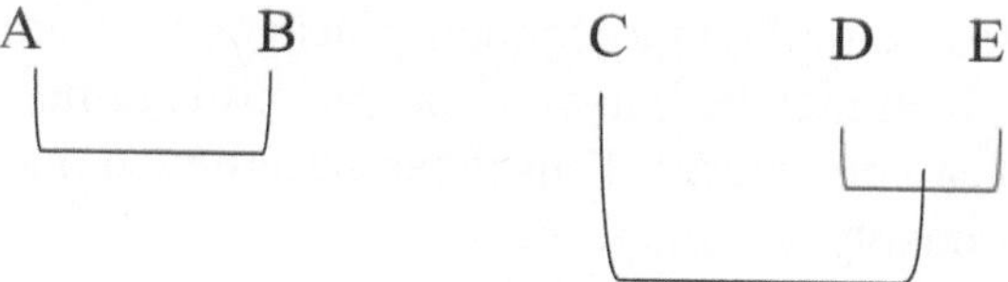

Species A has a distance of 3 to C:

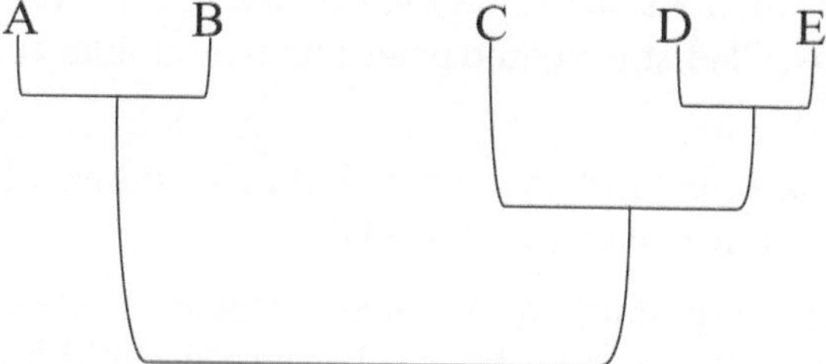

Each cluster formed contains lower order clusters (or species) and so the method is hierarchical and could be used as the basis of a conventional classification with Linnaean hierarchy. It was suggested that this method does not attempt to suggest phylogeny.

In 1960's Willi Hennig published a number of works giving a formulized method of estimating the phylogeny of a group of species. Soon computerised methods were derived from the work of Hennig and others, and they formed another technique called CLADISTIC method.

2. A Simple CLADISTIC Method

This method was derived by Hennig and it is based on the following assumptions:

1. That the direction of evolution of each character be known, eg. State '0' might be ancestral and state '1' derived (at a later stage in the evolution of our species group).
2. That evolution is an efficient process which means that only the minimum number of character state changes have taken place. This assumption is known as evolutionary PARSIMONY.
3. The reversal has not occurred, which means that when a character has changed from its ancestral to derived state, it will not subsequently change back to the ancestral state.

These assumptions appear over simple and biologically meaningless, but cladistics has some advantages over phenetic method. In formulating a classification, or simply identification, species with likely habits whose biology is unknown is taken into consideration. Therefore, a classification predicts data that are unknown. For example, in a phytophagous genus of ten species, the host plant of three of them may be known to be various legumes, and all related genera are largely legume feeders. Thus it can be assumed that the remaining seven species are probably legume feeders.

Cladistic methods appear to be the most predictive taxonomic techniques available, and hence they are good basis of classifications of applied entomology. Incidentally cladistics methods are scientific because predictable hypothesis form testable hypothesis. Cladistic method bears much resemblance to true phylogeny.

Firstly, the ancestral state of each character must be estimated and a hypothetical species is described by these states; it is called a GROUNDPLAN.

In these data set, the '0' state will be assumed to be **ancestral** and '1' state is assumed **derived.** The ancestral state must be known because it could be retained along many unrelated evolutionary lines. Therefore, only the sharing of **derived** states by species implies possible common ancestry.

The next stage is to form a table of the number of shared derived steps common to each species pair:

	A	**B**	**C**	**D**	**E**
A	-	5	2	1	1
B		-	2	1	1
C			-	1	1
D				-	2
E					-

A CLAD of species is formed with the above ground plan. Species A and B construct a hypothetical species called AB as their common ancestor. AB has a character in the derived state ('1') only if both A and B have maximum characters in derived state ('1'). A new data matrix is now formed without species A and B:

	1	2	3	4	5	6
AB	1	1	1	0	1	1
C	0	0	1	0	1	0
D	0	0	0	1	1	0
E	0	0	0	1	1	0

And the shared derived steps matrix is now:

	AB	**C**	**D**	**E**
AB	-	2	1	1
C		-	1	1
D			-	2
E				-

So another hypothetical species can be formed to join AB with C:

	1	**2**	**3**	**4**	**5**	**6**
ABC	0	0	1	0	1	0
D	0	0	0	1	1	0
E	0	0	0	1	1	0

And the shared derived steps matrix is now:

	ABC	**D**	**E**
ABC	-	1	1
D		-	2
E			-

Now a hypothetical species DE is formed from species D and E:

	1	**2**	**3**	**4**	**5**	**6**
ABC	0	0	1	0	1	0
DE	0	0	0	1	1	0

Finally, a cladogram can be drawn upon in which the derived steps are plotted as black squares:

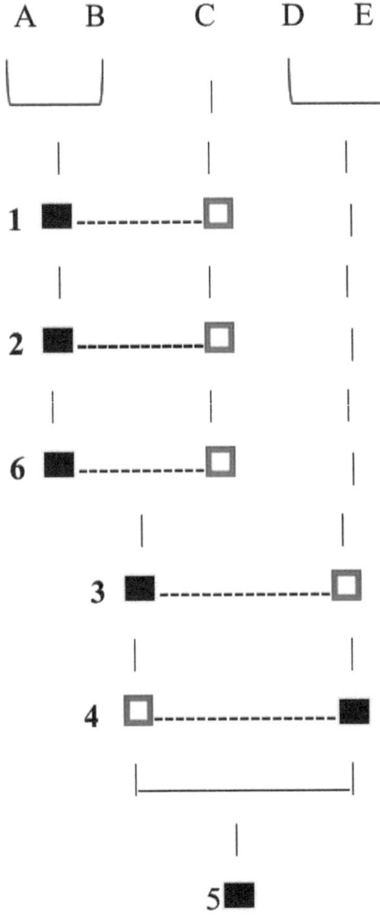

In the above cladogram derived characters 1,2,6 in AB are not shared with any other species. Characters 3,4 are shared with AB and C only and DE has got different characters 3 and 4, while 5th character is common in all. We can now discuss these techniques, with especial regard to the differences between them, and try them out by means of a computer simulation.

DENDROGRAM

Dendrogram is a diagram that shows the hierarchical relationship between objects. It is most commonly created as an output from hierarchical clustering. The main use of a dendrogram is to work out the best way to allocate objects to clusters.

It is a type of tree diagram showing relationships between similar sets of data. A dendrogram can be a column graph or a row graph. Some dendrograms are circular or fluid shaped but software will usually produce a row or column graph.

A basic graph has-

(a) A *clade,* which is the branch. Usually labelled with letters from left to right.

(b) Each clade has one or more leaves.

The clades are arranged according to how similar (or dissimilar) they are. Clads that are close to the same height are similar to each other; while with different height, the more dissimilarity.

11

Concept of Species and Kinds of Species

The Species Concept in Systematics

The modern biological or, more strictly, genetical concept of a species is a group of interbreeding organisms, genetically isolated from other such groups. The definition of the category species has long been one of the major problems of taxonomy. It is because of two reasons:

1. Because of the congruity of lingering typological concepts with the nature of evolving species.
2. Because evolutionary species may have changing characters and are only gradually differentiated in phylogeny.

For evolutionary study and descriptive purposes, many kinds of variation within species come in to consideration, but the only regular intra specific taxon is the subspecies. Subspecies are widely but not universally useful and their definitions yield to the same sort of criteria as for dividing other phylogentic continuance.

To the ancients and to the Scholastics, species were neither unchangeable nor having fixed boundaries.

- **Linnaeus** wrote his famous aphorism that there are as many species as were created in beginning.
- **Aquinas** stated strongly and clearly that new species are constantly being produced (but not by evolution). The above state aphorism by Linnaeus might have drawn Aquinas to his idea on species concept which was quite opposite to that of Linnaeus.
- Post Darwinian taxonomists observed that there are units in nature that have special evolutionary status not fully shared in the hierarchy. The units are frequently so distinct that they can be recognized without knowledge of their specific nature.
- There can be no one definition of the species applicable to all organisms and unambiguous in its application to every individual case. The two reasons are:

1. Member of a category around the level of species are not invariably separable in to groups by any absolute criterion.
2. Groups recognizable as species may get differ in population structures and mode of origin so that their category can not be adequately defined by the same evolutionary criteria for all cases.

The Genetical Species Concept

Mayr's definition (1940) has been most widely accepted. According to him, "Species are groups of actually or potentially interbreeding natural populations, which are reproductively isolated from other such groups."

Although often called the biological species concept by definition, that is strictly genetical concept of species among bi-parental organisms.

The genetical definition of species does not mention morphology which suggests the possibility that some good genetical species might not differ in anatomy. Therefore, pairs of anatomically close similar genetical species are called cryptic or sibling species, for example, in which genetical divergence is demonstrably great while anatomical divergence is hardly appreciable, such as the famous case of *Drosophila pseudoobsura* and *persimilis* (**Dobzhansky,** 1951).

Limitations of Mayr's Definition

- It does not operate in uniparental organisms.
- It is difficult to test the concept of 'potentially' for interbreeding.
- The third difficulty relates to the degree of isolation. A few taxonomists have insisted on absolute permanent isolation that is the impossibility of production of fertile hybrids.

The Evolutionary Species Concept

The evolutionary Species relates the genetical species directly to the evolutionary process that produce it. According to **Simpson** (1951):

"An evolutionary Species is a lineage (an ancestral sequence of population) evolving separately from others and with its own unitary evolutionary role and tendencies."

The definition makes it clear that two species may interbreed to some extent without losing their distinction in evolutionary roles and that is an important point for evolutionary taxonomy. The amount of interbreeding allowable by definition is then precisely as much as does not cause their roles to merge. The evolutionary concept is more readily related to paleontological sequences.

Simpson's Species Concept applies to groups of whole organisms. If the definition of morpho species literally means that morphology is taken into consideration, then classification would be one of characters and not of organisms.

Biological Species Concept

The biological species concept was introduced by Mayr. According to this concept species are groups of interbreeding natural populations that are reproductively isolated from other such groups (Mayr 1940). Thus, this species has three functions:

i) It forms a reproductive community i.e.; the individuals of a species recognise each other as potential mates for the purpose of reproduction. The species-specific genetic programme of every individual ensures intraspecific reproduction.

ii) It is an ecological unit regardless of the individuals composing it. It interacts as a unit with other species with which it shares the environment.

iii) It is also a genetical unit consisting of a large intercommunicating gene pool whereas the individual is holding some content of gene pool for a short period.

Nominalistic Species Concept

This concept is the concept of Occam and his followers, who believed that nature only produces individuals while species are the creation of man. In nature they lack definite existence. These concepts do not have any scientific basis. This concept was popular in France in 18th century but it is still in use, though very rarely.

Typological Species Concept

This concept manifests that the observed diversity of the universe reflects the existence of a limited number of underlying universals or types. Individuals are not related to each other but are only expressions of the same type. If two individuals or groups of individuals appear sufficiently different, they are different species. Variations of any individual is considered irrelevant and trivial phenomenon.

Karl Popper suggested essentialist species concept. According to it the species can be recognized by essential features or essential characters, and these are expressed in their morphology or species in a group of organisms. Characteristics such as colour, size, habitat etc. segregate them from all other organisms. It is therefore, also called **morphological species concept**.

Recognition Species Concept

Paterson (1985) and Lambert (1987) put forward this concept to replace the biological species concept. According to the concept a species is "that most inclusive population of individual biparental organisms which share a common fertilization system".

This concept did not receive support. Butlin (1987) stated that Paterson and his colleagues misinterpreted Dobzhansky and Mayr's original ideas.

Kinds of Species

Apart from true taxonomic species there are many kinds of species which pertain to ecological or evolutionary concepts. Such kinds of species may give clear understanding of a species. These are:

1. **Sympatric species:** Species that occupy the same geographical area.
2. **Allopatric species:** Those normally inhabiting completely different geographical areas.
3. **Sibling species:** Pairs or groups of similar or closely related species which are reproductively isolated but morphologically identical or similar looking.
4. **Polytypic species:** As defined by Huxley (1940), a species consisting of two or more subspecies.
5. **Monotypic species:** A species with no subspecies.
6. **Contemporaneous species:** Those species which occur at the same time level. These species indicate the number of phylogenetic lines or lineages occurring at a particular time.
7. **Transient species:** Species existing contemporaneously, as a cross section of the lineages or evolutionary species.
8. **Continental species:** Species living on large land masses, distinct from insular species.
9. **Insular species:** Species living on isolated islands which spread through dispersal methods other than overland migration.
10. **Incipient species:** These are geographical subspecies or other segregates, which if become isolated, will be distinct species.
11. **Morphospecies:** Those which establish by morphological similarity regardless of other considerations.
12. **Cosmopolitan species:** These are widely distributed species over the earth, in majority of zoogeographical regions.

13. **Non-dimensional species:** The taxonomic species lacking dimensions of space and time, applicable to only non-evolving animals and so not acceptable biologically. It is used during speciation.
14. **Pantropical species:** These are living or growing throughout the tropics. The term is ambiguous.
15. **Montane species:** Those which occur at high elevations on mountain ranges.
16. **Morpho-geographical species:** Described by Linnaeus, presently also recognized based on morphological and geographical data (basic species).
17. **Panmictic species:** Species characterized by random mating in which each sex is produced by either dioecious or monoecious individuals. Some progenies are the result of cross fertilization between different individuals.
18. **Apomictic species:** The species are mostly unisexual, there is no mating between different individuals. The individuals produce only ova. Some reproduce only asexually by budding or fission and have no functional sexual stage in their life history.
19. **Parapatric species:** Species have a narrow area of overlap. Such parapatric ranges are due to competition, combined with critical ecological boundaries.
20. **Paleospecies:** Also called Successional species, is a temporary species in a single evolutional line or lineage and integrate smoothly with each other.
21. **Palaeontological species:** A fossil species.
22. **Philapatric species:** Species which do not show any tendency to extend their range.
23. **Form species:** Fossil objects in group or isolated, such as fragments or isolated parts, which are not identified to any particular biological species.
24. **Paraspecies:** These are parataxa at the species level in palaeontology. Isolated parts of animals are named equally as species and genera although such fragment species or paraspecies may include objects belonging to several species. A new term parataxonomy includes study of parataxa, an artificial taxon used to describe a fossil based on morphology with no relation to genetic base.

Cain (1954) distinguished the following kinds of species:

1. **Taxonomic Species**: A general expression for any taxon that has been called a species and given a specific name available under the International Rules of Nomenclature.

2. **Morpho Species**: Established by morphological similarity regardless of other considerations.
3. **Palaeo Species**: Temporally successive species in a single lineage.
4. **Bio Species**: Means genetical species which is discussed above.
5. **Agamo Species:** Species of uniparental organisms.

Mayr (1942) distinguished two types of species-

i) **Allopatric Species:** Species occurring in different area. The genetical significance is that allopatric species do not actually have an opportunity to interbreed. They may even retain the potential for interbreeding while not infact interbreeding and undergoing considerable morphological and functional divergence.

ii) **Sympatric Species:** Species occurring at least in part at the same localities or overlapping populations. Sympatric species ordinarily remain such only if they interbreed within and not between species and do maintain quite separate roles.

Species in Uni-Parental Organisms (Agamospecies)

Most taxonomists (and animal geneticist) have tended to ignore uni parental organisms to make no attempts to define the category species with reference to them. **Dobzhansky** (1939) has insisted that there can not be a species category for uni parental organisms.

Meglitsch (1954) gave following definition to uni and bi parental species:

"The species, in the case of uni parental and bi parental organisms, may be visualized as natured population, evolving as unit in actuality, or retaining the capacity to evolve as a unit if artificial barriers are removed."

Evolution as unit, or the retention of a unitary role, is maintained by:

i) Community of inheritance.

ii) The capacity for genes to spread throughout the population.

iii) The inhibition of their spread to other populations.

All three factors occur in both uni and bi-parental populations.

In bi-parental organisms, the genes between populations are involved in interbreeding, which does not occur in uni- parental organisms. But it also involves movement of individuals from one population to other, production of offspring there, and their integration and continued reproduction within the second population. In the second population all the factors are equally relevant to uni parental population. Isolation in both case is promoted by failure to migrate, by lack of effective reproduction in the new environment or by failure of integration of the offspring.

Species in Paleontology

The data for classification of fossils are characteristically different from those in neonatology. Usually, no direct physiological or behavioural data are available. Ecological data, although available in significant amount are generally also much less complete than for recent animals. Large samples are less often obtainable and may not reach a conclusion.

12

Concept of Supraspecific and Infraspecific Catagories

Super Species

According to **Mayr** (1931, 1942) **Linsleg** and **Usinger** (1953),"A super Species is a monophyletic group of very closely related and largely or entirely allopatric Species".

Super species are, in other words, groups of populations that seem on other grounds (morphology, ecology etc.) to have passed beyond the point of potential interbreeding and to have acquired separate evolutionary roles. They are so by the more conclusive evidence of remaining separate when sympatric. It is to be assumed that they are still near the critical point of specification, near a definitive isolation. They are nascent species that will, if they survive, collectively form a subgenus or eventually a genus but have hardly get reached that degree of divergence.

They are not given special names; the Rules of Nomenclature make no provision for that usual designation and they are mostly known by the name of an included species and are in fact super species.

Sylvester - Bradley has proposed the term super species to be used in a regular taxonomic way for a taxon, that is a section of branching phylogeny larger than a species but smaller than a genus (or presumably, subgenus).

Thus Super species falls naturally into the hierarchic category sequence: sub species – species – super species – Sub genus and so on.

Infraspeific Category

Infraspecific category basically includes subspecies.

Subspecies

Subspecies is defined as "Geographically defined aggregates of local populations which differ taxonomically from other such subdivision of species".

Not more than one subspecies of any one polytypic species can exist in breeding condition in any one area. Adjacent subspecies can only interbreed or are potentially capable of doing so if separated by extrinsic barriers. It is now established that not only the species but subspecies also is an assemblage of populations. Therefore, a subspecies must be taxonomically different from other subspecies.

Many local populations showing characters separating themselves from other localities is the subspecies. They are taxa of a markedly different kind from species, and relatively few of them will ever become species although some achieve that status.

Characteristics of Subspecies

i) They are not little species, nor are they usually incipient or near species.

ii) Sub species do not express the geographic variation of the characters of a species and are only partially descriptive of that variation.

iii) They are formal taxonomic population units, usually arbitrary. The variation in those populations can express or fully describe phylogeny.

Ornithologists of 1844 and Schlegel (1954) began to recognize such geographical groups as taxa in classification and given them trinomials, adding a third name after the Linnaean specific binomial.

Geographical Race

The word race is currently used for mammals, birds and insects. Mostly, term subspecies and geographical race are used synonymously and interchangeably. The taxonomists term them as simply local populations. A subspecies is geographically localised and consequently a geographical race. Warm blooded and highly mobile animals, such as birds are rather independent of local environment, therefore, they are geographical races.

Ecological Race

No two localities are exactly identical with respect to their environment, therefore, the subspecies is also known as ecological race. Plants and many sedentary cold blooded animals are broadly exposed to the effects of local environmental conditions, and therefore, these subspecies are ecological races.

Geographical Subspecies

These are aggregates of population or synchronic infraspecific populations which are isolated macro-geographically during their mating time but whose respective members would crossbreed freely and normally if the populations were sympatric under natural conditions.

Geological Subspecies

Populations which function during different geological times have absolutely no chances of becoming synchronic. Hence, the probability of how freely and normally their members could interbreed is purely a speculation.

Ecological Subspecies

Aggregates of populations which are isolated micro-geographically but whose members would crossbreed rather freely and normally if the populations were to become micro- geographically sympatric under natural conditions. These occur in different niches, biotopes or their populations which require the use of special faunistic or topographic maps to indicate the exact micro-geographic habit of the concerned ecological species.

Temporal Subspecies

These are sympatric infraspecific populations or which are temporally isolated during mating season but whose members would crossbreed freely and normally if the populations were to become sympatric under natural condition.

Seasonal Subspecies

If two distinct sympatric populations or aggregates of populations within a given species are composed of individuals which mature at different stages during the same calendar (e.g. one in spring and other in rains) with no period of time during which reproductive forms of both demes coexist, then interbreeding can occur between the members of the respective populations.

Annual Subspecies

If the members of one distinctive population or aggregates of populations within a given species mature only during different year from those of another population of the same species, then the respective populations might be termed annual subspecies, if temporal isolation between their reproductive forms is complete.

Polytopic Subspecies

It is composed of widely separated population or heterogenous subspecies. When subspecies of a species differ in a single diagnostic character like colour, size, or pattern, it may happen that several different and somewhat widely separated populations independently acquire an identical phenotype population.

Intermediate Population

Intermediate populations are usually found in the area of contact of two well defined subspecies. Individuals of such intermediate populations may either be more or less uniform in character namely intermrdiate between the topotypical population of the two adjacent subspecies 'a' and 'b'; or this intermediate population may be a mixture of individuals, some of which resemble 'a' and others with 'b', while still others are intermediate.

Cline

Cline was proposed by **Huxley** (1939, 1940). These are a series of adjacent populations in which the gradual change of characters occures.

Local Population

Subspecies is the lowest taxonomic category. However, the subspecies are not homogenous but are composed of numerous local populations. These are differing slightly in gene frequencies and the mean values of various quantitative characters.

Deme

Morphologically homologous group of organisms which are either from a single locally or from a single kind of habitat. It is the starting point in all classifications. **Mayr** (1953, 1969) called it PHENA and **Simpson** (1961) called it DEMES. It is a group of individual animals of one species or subspecies so localised that they are in a more or less frequent contact with each other. They are unispecific members of a single community in the most limited sense.

Variety

A term originally applied indiscriminately to various kinds of infraspecific forms, individuals, as well as populations. In modern usage, it is limited to discontinuous variants within a single interbreeding population (subspecies, infraspecific forms). Under the typological principles, each species had a fixed pattern and anything that is different was named as variety.

Types

Whenever a new species or other group is described, a describer is supposed to designate a type, which is used as a reference, if there is ever any question what that species or group includes.

The type of a species or subspecies is a specimen, the type of a genus or subgenus is a species and the type of a family or sub family is a genus.

Biotypes

A population or group of individuals composed of a single genotype.

Genotype

In nomenclature, the type species of a genus (ef. Type species); in genetics, the class in which an individual falls on the basis of its genetic constitution, without regard to visible characters (ef. Phenotype).

Holotype

Species selected by the author of the species from the type species.

Paratype

Each specimen of a type series other than the holotype. It may happen that an author will describe a new species from a single specimen the holotype, and then discusses variability in the species by referring to other specimens that he has designated as paratype. The holotype and paratype(s) are collected from the same locality but this does not have to be so. Authors often deposit paratypes in different museums, in case the holotype and its associated types are destroyed. A paratype has a junior status to the holotype so if later they are found to be different species, the paratype is discarded as a type of the taxon and the holotype retained.

Strain

A biological strain of an organism, morphologically indistinguishable from other members of its species but exhibiting distinctive physiological characteristics; particularly in regard to its ability to successfully utilize pest-resistant host organisms or to act as an effective beneficial species.

Syntype

A series of specimens designated by the author of first species at the time of establishing species.

Lectotype

A specimen selected by reviser from among Syntypes.

Paralectotype

The remaining specimens of a syntype series after a lactotype has been designated.

Neotype

In case Holo, Syn or Lecto types are lost or destroyed, a new specimen is designated.

Allotype

It is a paratype of opposite sex to the holotype. If the opposite sex is described at a later date to the original naming of the taxon, one cannot at this stage validly designate an allotype.

Copytype

This term is no longer used indicating a syntype or paratype.

If the existing type of a species is useless for distinguishing the species (making the name a *nomen dubium*), an author can ask the commission to allow him to designate a neotype instead.

13

Procedures in Identification

Identification of the insects is the integral part of systematics. It utilises all taxonomic procedures. The collections in the world are the accumulation of past taxonomists and are stored for the future identifications. Every species name is based on a published description or figure and usually on the type specimen. Identification is the association of other specimens with the appropriate description or with type specimen.

The steps in identification are as follows:

1. Preliminary key to orders and families
2. Key to genera and species if recent monographs or faunal works are available
3. Reference to the most recent catalogues
4. Reference to current bibliographies for literature published subsequent to the most recent catalogues
5. Reference to original descriptions
6. Comparison with the authentic described specimens or with types.

1. Preliminary key to Orders and Families

For a beginner this step is very important and for this, simple keys given in general text books or hand books can be used. In addition to the general keys, a bibliography to the most important regional works on each group of insects can be taken into consideration.

2. Key to Genera and Species

The specimen is thoroughly run through the keys, the description of the appropriate species is checked character by character, and if illustration is available, it is compared with it along with the geographical distribution. If all these points coincide, the identification is considered as tentatively made.

3. Reference to Recent Catalogue

Most recent catalogues are consulted in case a monograph is not available, to get literature citations to the descriptions of all species. They may also give complete biographies, under each genus and species, lists of synonyms and geographic distribution. Identification is greatly helpful by a good catalogue, because it brings together the most significant published reference in the group and guides to reach the most likely species in the territory from which the specimen was collected.

4. Reference to Current Bibliographies

In case the recent catalogues are not available, an annual bibliography of the literature in systematic zoology can be consulted as mostly the catalogues are out of date soon after they are published. This reference book is called the *Zoological Record* in which each scientific name is given, together with a reference to the place of publication and the type of locality.

The Zoological Record is published by Zoological Society of London in cooperation with British Museum (Natural History) and the CAB International Institute of Entomology. Biological Abstracts (1926 to update) is an important source of recent literature. Its section, Systematic Zoology, contains abstracts of taxonomic papers which are not immediately available elsewhere.

5. Reference to Original Descriptions

Descriptions are the foundation of taxonomy, since the printed work is in-destructive. Types may be lost and the original author may be available for only a brief span of years. Illustrations are equally valuable in case of birds and butterflies.

Keys are good source of identification but original or more recent authoritative descriptions be used in case the specimen in question is not covered in the key.

6. Comparison with Types and Other Authentically Determined Specimens

In case the identification is not determined with the existing literature, comparison of specimen is made with whatever series of specimens is available. Care should be taken not to rely exclusively on comparison with the supposed specimens. Even authoritative collections may contain wrong identifications. Type specimens are the most authentic of all but ideally for a monographic study all type specimens should be re-examined.

To identify the subspecies, the type specimen is not always required. On the other hand, a series of specimens from the locality of type ('topo- typical specimens) is required to provide information on the characters and variability of subspecies.

Determination Labels

As soon as the identification is complete, each specimen should be labelled. The label should narrate the scientific (generic, species, subspecies) name and author; name of the person who determined, place and the year in which the identification was made.

Tentative or doubtful identifications should always be clearly indicated as such by means of a question mark.

Taxonomic Characters

The aim of modern taxonomy is not only to describe, identify and arrange organisms in convenient categories, but also to understand their evolutionary histories and mechanisms. Earlier approaches were mainly based exclusively on observed characters without understanding the interspecific differences. Many of the species are therefore, known as single or a few specimens. Presently great attention is given to subgrouping of the species like subspecies and populations. The old morphological species is now called a biological species which also includes ecological, genetical, biochemical and other characters. All these new approaches have contributed a lot in explaining the true structure of the species and its evolutionary position. The current approaches in taxonomy are as follows:

1. Morphological Characters

i. General External Morphology

The external morphology provides a primary source of taxonomic characters. Animals with an external skeleton, for example, arthropods, molluscs etc. are the most important and useful characters of external structural characters. They range from superficial characters as plumage and pelage characters to highly conservative and phylogenetically significant suture and sclerites of the arthropods body.

ii. Genetalia Structure

The reproductive isolation is a must at the species level, therefore, difference in genetalia have been employed in many groups as last attempt to ascertain a species. It has been suggested that a look – and – key relationship exists as regards the copulatory structures of the male and females of those species with sclerotized genetalia. On the other hand, genital characters have been found to vary in the same manner as other characters. In general, genetalia differences must be evaluated just like other characters. In groups where genetalia significance has been proved, it was very useful, because, genetalia structures are the first to change in the course of speciation.

iii. Internal Anatomy

In higher animals, anatomy provides an abundant source of taxonomic characters. Generally, external morphology characters are in inverse ratio to the abundance and usefulness of internal morphology. In many vertebrates selected portions of internal skeleton are routinely preserved and used in identification. Palaeontologists deal with hard parts of many skeleton characters in groups of animals with internal skeleton.

iv. Embryology

Comparative embryology offers taxonomic characters of great phylogenetic significance. Thus, cleavage patterns, gastrulation and other embryological phenomena may be characteristic for whole phyla and help in understanding of higher categories. In insects, the total (holoblastic) cleavage of the collembola (springtails) emphasises the wide gap which seperates this group from the other Apterygota (primitively wingless insects) and the pterygota, in spite of the appearance of this cleavage type in a few highly specialised parasitic Hymenoptera.

v. Karyology

Karyology and other cytological characters deal with the variation in chromosomal structure, which are useful in identification. The simplest cytological characters are chromosome number. This is determined by a relatively simple technique involving the crushing or smearing the testes on a slide. Chromosome numbers have been recorded for a number of animals, and these studies have been used as evidence of phylogenetic relationships with others.

Dobzhansky, Patterson and Sturtevant, as well as several other authors have made substantial contributions in recent years to our knowledge of chromosomal variation in Drosophila. Some closely related as *Drosophila pseudoobscura* and *D. persimilis* are diagnosed more accurately by their chromosomal configuration than by other features. In a study of Finnish bug of family Lygaeidae, all the genera and nearly 56 cytological investigated species could be identified by their chromosomes alone. A good summary of this field has been given by White 1954. Some of these cytological differences interfere with chromosomal pairing and thus serve as isolating mechanism.

Such studies are useful with a good knowledge of cytology. The number of chromosomes may be different in close relative (due to the joining of two chromosomes after the loss of a kinetochore); genetically inert chromosome sections also are easily lost. Two species with superficial identical chromosomes may be much more different geneticaly than others with various gross chromosomal differences. The chromosomal polymorphism in species of *Drosophila* and others supply excellent evidence for this.

2. Physiological Characters

Physiological characters can be

a. Metabolic factors
b. Serological, protein, and other biochemical differences
c. Body secretions
d. Genic sterility factors

Physiological processes are considered as structures that are product of growth processes. These are difficult to analyse. Most of the physiological processes are regulated by enzymes and other micro-molecules and thus are not separable from biochemical characters. Physiological characters generally mean growth constants, temperature tolerance and other various processes. These characters are abundant in species differences but they are not present in preserved material and usually require special apparatus for their study. Therefore, they are rarely used by taxonomists.

3. Biochemical Characters

Biochemical characters such as macromolecules and metabolic processes are found in earliest living organisms such as Procaryotes as well as in highest animals. At taxonomic level this is exploited to make identification. Serology provides the widely used method of comparing proteins. This method is based on the principle that the proteins of one organism will react more strongly with antibodies to the proteins of a closely related organism than to those of one more distantly related.

This method encounters many technical difficulties and though it was used for a long time, it has not proved to clarify the otherwise ambiguous cases. This technique has been revived with the quantitative study of antigen reactions. The study of blood group genes (Immunogenetics) is now used extensively in the study of primates.

4. Cytological Characters

i. Chromosomes

Chromosomal patterns have been used for plants far longer than the animals. Improvements in cytological techniques during the past fifty years now permits chromosomal studies even in so called birds and lepidopterans which have small chromosomes and high chromosomal numbers. It caused great difficulties. Diptera with giant chromosomes and Orthoptera are most suitable for chromosomal studies. Chromosomal patterns have been studied for Drosophila, chironomids, homopterans etc.

Chromosomes are particularly useful on two different levels.

1. They aid in the comparison of closely related sibling species. These can be similar in external morphology but different in chromosomal patterns.
2. Chromosomal patterns are of extreme importance in establishing phyletic lines. Most chromosomal changes are unique which are then characteristic for all descendants of the ancestral population in which the new pattern first became established. Patterns of replacement often indicate whether or not two organisms belong to the same phyletic line.

Changes in sex determination, in all sorts of rearrangements of the chromosomes and centromeres, in fusions, fissions or translocations, in the acquisition of supernumeraries, etc., often give unequivocal clues to relationship. For instance, the similarity in the spermatogenesis of Mallophaga and Anoplura (true lice) strongly supports the belief in a close relationship of these taxa.

The chromosomes represent the genetic material but the amount of chromosomal change does not reflect the amount of genetic change. Close relatives may show considerable rearrangement. Many species are polymorphic for various types of chromosomal rearrangements. Cases are known in which a considerable degree of genic change is not or only lightly reflected in the chromosomal pattern, Example- the Hawaiian *Drosophila*. Techniques of DNA matching are now being developed which hold much promise.

ii. Chemical Component

Specific chemical components and macromolecules have been used in taxonomy recently. Various methods of electrophoresis reveal the molecular composition of complex proteins. Paper electrophoresis, which was used in the earlier studies, has been largely replaced by newer techniques that permit a much finer resolution, but still newer techniques are continually being introduced.

iii. Haemoglobin

Another approach is to consider a single complex molecule for instance the haemoglobin, of one species, and compare its amino acid composition with that of closely or more distantly related species.

When using taxonomic characters to draw inferences on classification, one must always balance the potentially conflicting information derived from different characters. The minute differences between evolutionary phenomena at the molecular and at the organismic level should also be considered.

5. Parasitological Characters

Parasites are important in contributing to our knowledge of the relationship of higher taxa. Parasites evolve together with their hosts and are in some cases more conservative than their hosts. Unfortunately, they shift to new hosts more frequently and the evidence on parasites is evaluated very carefully. It was noticed in several instances that sibling species were discovered because their parasites were different. A new species of termites was discovered because its nests contain a different set of Termitophile staphylinid beetles than those of a previously known species. In another example the flamingos (*Phoenicopteri*) exhibit characteristics which they share with both storks and geese.

In case of bird lice derived from a common ancestor, it is expected that they are related, but slightly different parasites in the two orders of birds have been noticed. Another example of the most primitive tribes of the coccids (Steingeliini, etc.) which have no symbionts, but once a coccid taxon has acquired them, this symbiont with highly specific adaptations, will be found in the derived phyletic lines of coccids. The same is true for the symbionts of other groups of insects. The protozoan faunas in the intestines of termites evolved together with their hosts and are potentially useful indicators of relationship in cases of ambiguity in the termite classification.

6. Geographical Characters

Geographical characters are among the most useful tools for clarifying ambiguous taxonomic species. There seems to be a correlation with geographical or associated ecological features with the species identification. In taxonomy two kinds of geographical characters are useful:

i) General biographic patterns, which are especially useful in the arrangement and interpretation of higher categories.

ii) The allopatric sympatric relationship which is most helpful in determining whether or not two populations are conspecific.

i) General Biographic Patterns

The broad geographic patters are determined by the study of distribution of large numbers of groups of plant and animals. The world has been divided by biographers into various realms, regions, provinces, sub provinces etc. depending upon generalised comparison of faunas and floras. Depending on the group, they may be expanding or retreating and thus are more useful to refer them as faunas or floras or biotas rather than zones or areas. In identification, taxonomist should have a good knowledge of the past geological history of the regions in which such biotas' centre exist and their relationship with the faunas and floras.

For example, the mammals of South America are either not related to those of Africa or, if of common ancestry have presumably reached South America by way of North America. The hystricomorphic which are close to the African porcupines, appeared to be exceptional, the history of which was inexplicable in view of absence of such forms in the early Tartiary period of North America. A close examination of these porcupines and their relatives revealed that the porcupines of South America and Africa are of independent origin. Distributional differences have shed light on taxonomic relationship in many other instances.

ii) Sympatric – Allopatric Relationship of Populations

This method is the most useful method to ascertain whether or not a distinct species exists in two populations. If a number of species or a ring of species, which differ from its neighbours, the forms are said to be allopatric. Such a distributional pattern in closely related forms is generally subspecies. On the other hand, if these forms do not integrate, they are said to be sympatric. This distributional pattern indicates that these forms have evolved into a full species, due to sympatric coexistence without interbreeding. This forms the basis of species concept.

14

Kinds of Taxonomic Keys

Taxonomic keys are devices that make identification of taxa easy and enjoyable. A key is presentation of appropriate diagnostic characters in a series of attractive choices to facilitate identification of a specimen.

Key characters are the characters that are easily perceived of very low variability, usually present in preserved material. They are useful as convenient labels for distinguishing taxa by the process of classification. Many taxonomic characters such as chemical, chromosomal, physiological and behavioural may have high value of identification because they are inaccessible in preserved material. Ideal key characters apply equally to all individuals of the population. They are absolute, relatively constant and external so that they can be observed directly without special equipment.

Different Types of Keys

1. **Dichotomous key:** This is the most common key where each couplet has two alternatives. This may be of two types:

a) Simple non- bracket key.

b) Simple bracket key.

a) Simple non- bracket key

This key is most often used. The couplets are easy alternatives for ready comparison. The specimens to be identified are run through a number of keys. It can be used from backwards but is slight difficult if run with backward clues.

Example:

1. Six legged.. 2

   - Eight legged 3

2. Two wings ....A

   - Four wings..B

3. Hard bodied C

   - Soft bodied .... D

b) Simple bracket key

This key is similar to non- bracket key but mostly differ in use of brackets in parenthesis after the main number mentioned for the key couplets. When made properly, it is more easy and quick to run through this key forward and backwards. This is the best key in fulfilling its identification purpose.

The main disadvantage in both these types is that the relationships of the division is not apparent as in case of indented key. This is a commonly used key by many taxonomists.

Example:

1. Six legged 2
   - Eight legged 3
2. (1) Two wings .... A
   - Four wings .... B
3. (1) Hard bodied C
   - Soft bodied .... D

2. Indented key

The couplets are indented from the left hand margin of the page, in such a way as to show their importance. Thus the two or more members of the primary couplets are near the left hand margin, while the secondary couplet is indented after leaving four or five species; the tertiary couplet with equal number of spaces beyond the secondary couplet and so on. This type of key is advantageous in the sense that the relationship of various divisions is quite visible to the eye and can be used in reverse also. When the key is short, it works but in long keys, the alternatives get widely separated and take more spaces. Thus, it serves good purpose for keys to higher taxa or comparative key.

It was once very popular but now not much used as it takes a lot of space. Indented key, although, has the advantage that the relationship of the various divisions is apparent to eye but it is long key. However, a modified indented key is more popular among European and Russian taxonomists which resembles dichotomous key.

Example:

A. Six legged
   - a) Two wings .... A
   - b) Four wings .... B

B. Eight legged

a) 4-7 cm long C

b) 8 cm long... D

3. Pictorial key

Pictorial keys are designed for special purpose of field identification by non-scientists. They can identify the commonly occurring species with the help of characters together with their figures in a comparative manner. There are several such keys for various groups of animals. Such keys are given in detail in this chapter.

4. Interactive key

All the keys are dichotomous and based on a series of choices.

Example:

Key to unknown specimens

1. Body covered with hard wings.. 2

   - Body covered with soft wings.. 3

2. Head prognathous, hard leathery, chitinous coloured wings .... beetle

   - Head hypognathous or opisthognathous 4

3. Wings covered with pigmented scales, clubbed antenna .... moth

   - Wing membranous, forewing larger with veins and cross veins distinct .... wasp

4. Head hypognathous forming right angle to body axis, chewing and biting type mouth parts, fore wing leathery .... hopper

   - Head opisthognathous, turned backward against the body axis, mouth parts piercing and sucking type. Forewing half leathery half membranous .... bug

Πιχτοριαλ Κεψ το τηε Ορδερσ οφ Ινσεχτα

S.No.	Characters of couplet	Figure
1	Winged insects ....2 - Wingless or vestigial winged insects 23	
2.	With only one pair of wings ....3 - With two pairs of wings 5	
3.	Wings net veined; haltares absent .... certain mayflies: Ephimeroptera - Wings not net veined; haltares present 4	
4.	Wings with highly reduced venation; caudal filaments usually present; delicate insects .... male mealy bugs and scale insects: Homoptera - Wings with longitudinal and a few cross veins; no caudalfilaments (Fig. 1) ...... flies, mosquitoes, etc: Diptera	Fig. 1
5.	Forewings horny, without veins, meeting in a straight line over middle of body and usually concealing membranous hind wings (certain forms have the hindwings vestigial or absent)...6 - Forewings not as above...7	
6.	Abdomen provided with forceps like cerci at posterior end (Fig. 2) .... Earwigs: Dermaptera - Abdomen without forceps like cerci (Fig. 3) .... beetles, weevils: Coleoptera	Fig. 2 Fig. 3
7.	Two pairs of wings unlike in structure.............................8 - Two pairs of wings similar in structure.......................10	
8.	Forewings reduced to slender club-shaped appendages hindwings folded fan like at rest (Fig. 4) ....twisted winged insects: Strepsiptera - Forewings not as above..9	Fig. 4

9.	Forewings thick and leathery at base and at tip; mouth parts form a sucking beak (Fig. 5) ....bugs: Hemiptera - Forewings thick and leathery throughout; chewing mouth parts (Fig. 6) .... grass hoppers, cockroaches, etc.: Orthoptera	 Fig. 5 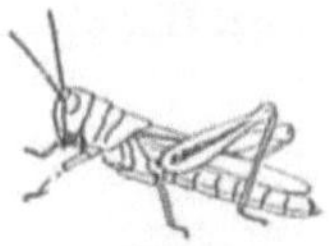 Fig. 6
10.	Wings partially or more often entirely covered by microscopic scales (Fig. 7) ....moths, butterflies: Lepidoptera - Wing transparent or covered with fine hairs................11	 Fig. 7
11.	Wings very narrow and fringed with long hairs, small slender – bodied insects (Fig. 8) ......thrips: Thysanoptera - Wing not as above..12	 Fig. 8
12.	Mouthparts a piercing.- sucking beak arising from the rear of the head near the first pair of legs (Fig. 9) .... aphids leaf hoppers etc.: Homoptera - Mouthparts not a piercing sucking beak, normally situated at the front of the head....................................13	 Fig. 9
13.	Antennae small and bristle like....................................14 - Antennae conspicuous and of many forms.................15	
14.	Fore and hind wings nearly equal in size; tip of abdomen without terminal filaments (Fig. 10).................................. .... dragonflies, damselflies: Odonata - Forewings much larger than hind wings; tip of abdomen with 2-3 long terminal filaments (Fig. 11)......................... .... Mayflies: Ephemeroptera	 Fig. 10 
15.	Wing with many veins and cross veins ....16 - Wings with few veins and cross veins 20	Fig. 11
16.	Hind tarsi with fewer than five segments......................17 - Hind tarsi with five segments.......................................18	

17.	Tarsi three segmented; hind wings as large as or larger (wider) than forewings (Fig. 12)..... stoneflies: Plecoptera - Tarsi four segmented; forewings and hindwings of equal size (Fig. 13) .... termites: Isoptera	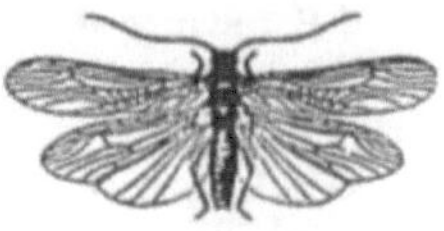 Fig. 12  Fig. 13
18.	Head prolonged into a beak (Fig. 14) .... scorpionflies: Mecoptera - Head not prolonged into a beak..................................19	 Fig. 14
19.	Wings covered with fine hair (Fig. 15)............................. .... caddisflies: Trichoptera - Wings transparent not covered with hair (Fig. 16) .... lacewings: Neuroptera	 Fig. 15 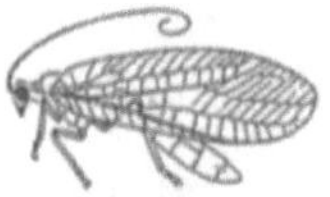 Fig. 16
20.	Tarsi two or three segmented; wings approximately equal in size ....21 - Tarsi usually five segmented; forewings larger than hindwings (Fig. 17) ...... ants, bees, wasps: Hymenoptera	 Fig. 17
21.	Basal tarsal segment of foreleg greatly enlarged (Fig. 18) .. .. web spinner: Embioptera - Basal tarsal segments not enlarged 22	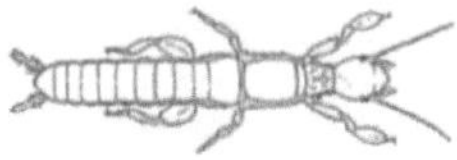 Fig. 18
22.	Cerci present; body less than 3 mm long (Fig. 19) .... zoraptarans: Zoraptera - Cerci absent; body 3 mm long or longer .... bark lice: Psocoptera	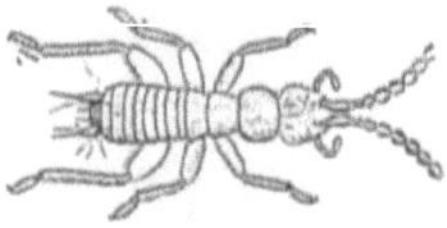 Fig. 19
23.	Abdomen composed of six or fewer segments with ventral spring apparatus (Fig. 20) .... sprigtails: Collembola - Abdomen with more than six segments and no spring 24	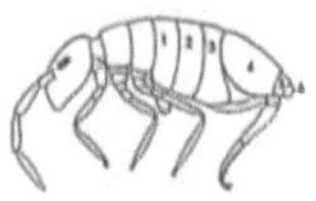 Fig. 20

24.	Abdomen segments 1-3 each with a pair of small ventral appendages; antennae, eyes and cerci absent; minute and rare (Fig. 21) .... telsontail: Protura -Abdomen and appendages not as above 25	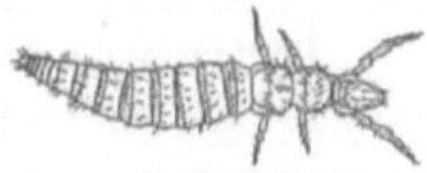 Fig. 21
25.	Abdomen with 2-3 long terminal appendages or pair of forcep like cerci, segments 2-7 may each have a pair of small ventral leg like appendages (Fig. 22) .... bristletails: Thysanura - Abdomen without terminal filaments or ventral appendages 26	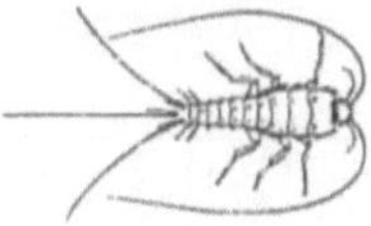 Fig. 22
26.	Mouth parts fitted for chewing.................................... 27 - Mouth parts fitted for piercing, lapping or sucking, sometime concealed...31	
27.	Louse like insects ....28 - Insect not louse like, various forms 29	
28.	Antennae with 5 or less segments (Fig. 23) ....lice: Mallophaga - Antennae with more than 5 segments (Fig. 24) ....booklice: Psocoptera	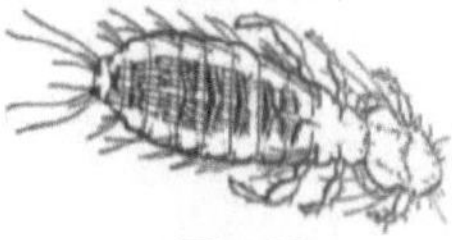 Fig. 23  Fig. 24
29.	Abdomen constricted at base ....ants, wasps: Hymenoptera - Abdomen not constricted at base 30	
30.	Body very slender and linear, hind legs modified for jumping, or body oval and flattened .... grasshopper, cockroach: Orthoptera - Body or legs not as above, body antlike but abdomen broadly joined to the thorax...................termites: Isoptera	
31.	Tarsi with 5 segments..32 - Tarsi with fewer than 5 segments...............................34	
32.	Body highly compressed laterally (Fig. 25) .... fleas: Siphonaptera - Body not strongly compressed laterally......................33	 Fig. 25
33.	Abdomen not distinctly segmented, covered with hairs ....sheep ked, other flies: Diptera -Abdomen distinctly segmented covered with scale .... females of bagworms, tussock moth: Lepidoptera	

34.	Last tarsal segment a bladder like organ, without well-developed claws .... thrips: Thysanoptera - Last tarsal segments with one or two claws 35	
35.	Louse like; sucking beak not evident (Fig. 26).................. ....sucking lice: Anoplura - Insect not louse like, sucking beak evident.................36	Fig. 26
36.	Beak arising from front of head .... bedbugs, water striders: Heteroptera: Hemiptera - Beak arising from rear of head (Fig. 9) .... aphids, scale insects etc.: Homoptera: Hemiptera	

Key to Major Economic Families of Orthoptera

S.No.	Characters of couplet	Figure
1.	Antennae much longer than pronotum; tympanum in front tibiae (Fig. 27)...2 - Tympanum about as long as pronotum; tympanum present on first abdominal tergum (Fig. 28) .... Grasshopper: Acrididae	Fig. 27 Fig. 28
2.	Tarsi four segmented; ovipositor usually sword like or sickle like (long horn grsshoppers) (Fig 29) .... katydids and others: Tettigonidae - Tarsi three segmented; ovipositor spear shaped or awl shaped (Fig. 30) ....crickets: Gryllidae	Fig. 29 Fig. 30

Key to Major Economic Families of Hemiptera: Suborder Homoptera

S.No.	Characters of couplet	Figure
1.	Antennae setaceous (bristle like) (Fig. 31)............2 - Antennae filiform (thread like) (Fig. 32) or rudimentary..4	Fig. 31 Fig. 32
2.	Pronotum extending backward over abdomen (Fig. 33) ....tree hoppers: Membracidae - Pronotum not extending backward over abdomen...3	Fig. 33
3.	Hind tibiae with one or more rows of spines (Fig. 34) .... leafhopper: Cicadellidae - Hind tibiae with one or two stout spines and usually a circlet of spines at apex (Fig. 35) .... spittle bugs: Cercopidae	Fig. 34 Fig. 35
4.	Tarsi two- segmented and with two claws...........5 - Tarsi one segmented, with single claw (Coccoidia)... 8	
5.	Hind femora large for jumping (Fig. 36); antennae with 5-10 segments (usually 10) segments .... Psyllids: Psyllidae - Hind femora not enlarged for jumping; antennae with 3-7 segments..6	Fig. 36
6.	Wings opaque, usually covered with white powdery wax (Fig. 37) ..... white flies: Aleyrodidae - Wings transparent when present.......................7	Fig. 37

7.	Cornicles (pair of tubules on top rear of abdomen) usually present and conspicuous (Fig. 38); wing venation not highly reduced.......aphids: Aphididae - Cornicles not present; wing venation highly reduced (Fig. 39) ....bark aphids, gall aphid and phylloxerans: Phylloxeridae	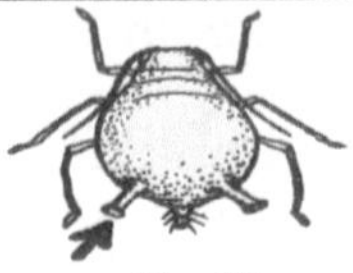 Fig. 38  Fig. 39
8.	Body hidden by waxy or scale like covering; sessile during most of life..................................9 - Body covered with powdery wax; mobile throughout life ....mealy bugs: Pseudococcidae	
9.	Body covered with hardened shell formed from wax; shed skins and fibrous material easily removable; females without posterior end cleft .... Armored scales: Diaspididae - Body covered with soft wax not easily removable; if not covered with soft wax, then females with hard, smooth, often greatly convex, endoskeleton with posterior end cleft ....soft scales: Coccidae	

Key to Major Economic Families of Hemiptera: Suborder Heteroptera

S.No.	Characters of couplet	Figure
1.	Antennae shorter than head; concealed in grooves beneath the eyes... ....(short horned bugs: most of the bugs) - Antennae longer than head and plainly visible from above (long horned bugs: terrestrial bugs and water striders)..2	
2.	Forewings reduced to short pads; body flattened and modified for actoparasitic habit.................................bed bugs: Cimicidae - Forewings not reduced to short pads; body not flattened and modified for ectoparasitic habit.......3	
3.	Membranous portion of forewing with two closed cells (Fig. 40)..........................plant bugs: Miridae - Membranous portion of forewing without two closed cells...4	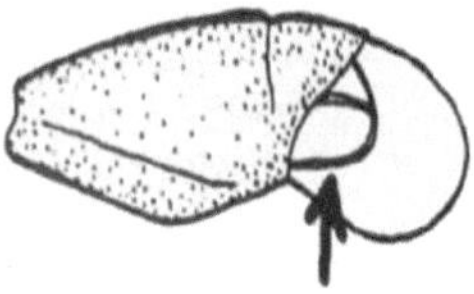 Fig. 40

4.	Membrane with row of small cells around margin (Fig. 41)....................damsel bugs: Nabidae - Membrane without row of small cells around margins..5	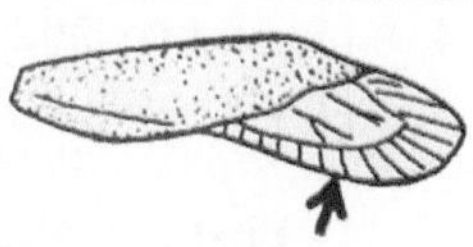 Fig. 41
5.	Membrane with 4-5 open veins (Fig. 42)............... ....chinch bug: Lygaeidae - Membrane without 4-5 open veins (may be many)..6	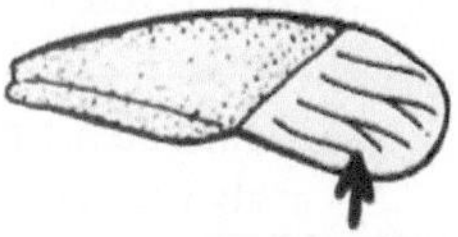 Fig. 42
6.	Membrane with many branched veins and cells but without numerous longitudinal veins (Fig.43) ....Red bugs: Phyrrhocoridae - Membranes without many branched veins and cells but with numerous longitudinal veins (Fig. 44)...7	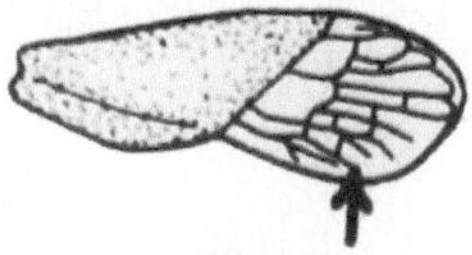 Fig. 43 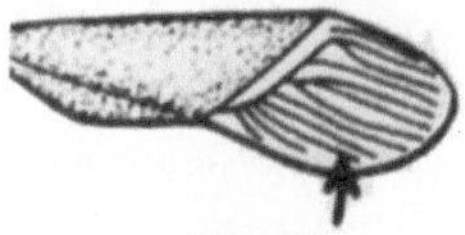 Fig. 44
7.	Antennae four segmented scutellum usually not large...................................squash bugs: Coreidae - Antennae five- segmented; scutellum very large (Fig. 45)...........................stink bug: Pentatomidae	 Fig. 45

Key to Major Economic Families of Coleoptera

S.No.	Characters of couplet	Figure
1.	Head not prolonged into a snout; gular suture double (on underside of head) ....2 - Head usually prolonged into a snout; gular sutures fused or lacking...13	
2.	First abdominal sternum divided by hind coxae (Fig. 46)......................ground beetles: Carabidae - First abdominal sternum not divided by hind coxae 3	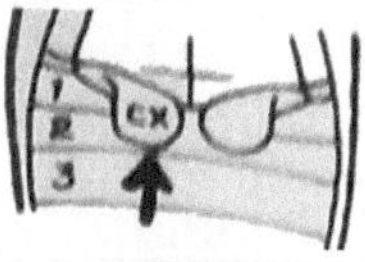 Fig. 46
3.	Click mechanism (prosternal spine fitting into groove in mesosternum) present (Fig. 47)..............click beetles, wire worms: Elataridae - Click mechanism not present..........................4	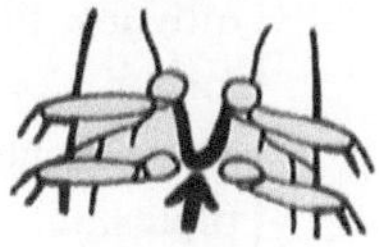 Fig. 47

4.	First two abdominal sternum fused (Fig. 48); body usually metallic .... flat headed wood borers: Buperestidae - First two abdominal segments not fused; body not usually metallic...5	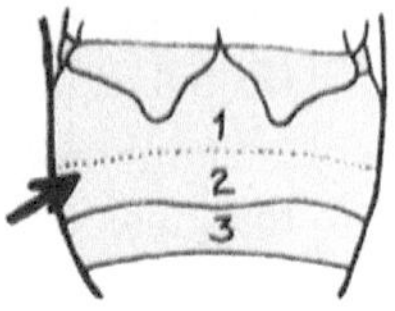 Fig. 48
5.	Hind coxae dilated and grooved not hairy or scaly beetles...6 - Hind coxae not dilated and grooved; not hairy or scaly beetles...7	
6.	Head concealed from above; front coxal cavities open behind (Fig. 49)..carpet beetles: Dermestidae - Head not concealed from above; front coxal cavities closed behind (Fig. 50).. fruit worm beetles: Byturidae	 Fig. 49 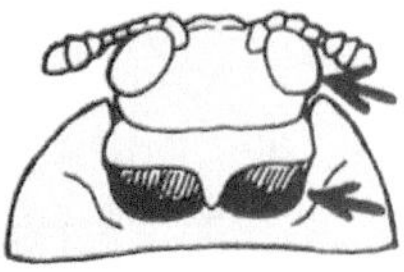 Fig. 50
7.	Tarsi usually three segmented (Third segment minute and fused to base of fourth (Fig.51); body almost hemispherical...lady bird beetle: Coccinellidae - Tarsi not apparently three segmented; body not almost hemispherical.......................................8	 Fig. 51
8.	Body highly flattened and narrow (Fig. 52)...flat bark beetles: Cucujidae - Body not highly flattened and narrow..............9	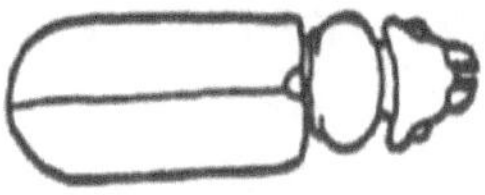 Fig. 52
9.	Tarsal formula 5-5-4 (Fig. 53)...................................Darkling beetles, flour beetles: Tenebrionidae - Tarsal formula not 5-5-4.................................10	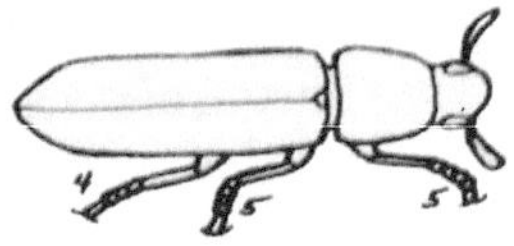 Fig. 53
10.	Antenna lamellate with club formed from movable plates (Fig. 54)...White grubs : Scarabaeidae - Antennae without club formed from movable plates..11	 Fig. 54

11.	Elytra short, exposing tip of abdomen; pronotum greatly narrowed anteriorly (Fig. 55)..pulse beetles: Bruchidae - Elytra not short; pronotum not generally narrowed...12	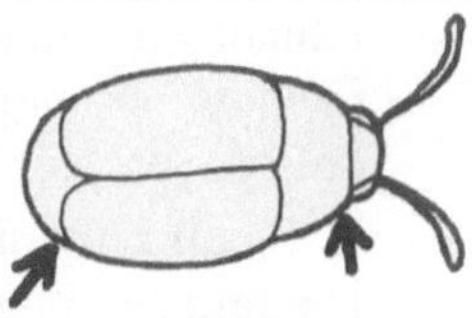 Fig. 55
12.	Antennae usually less than ½ length of body; usually small, rounded and brightly coloured. .... leaf beetles: Chrysomellidae - Antennae at least ½ length of body, often longer; small to large, elongate and with or without colour patterns rounded.........wood bores: Cerambycidae	
13.	Head prolonged into a definite snout (Fig. 56); small to large beetles of varying shapes .... weevils: Curculionidae - Head slightly prolonged into an obscure snout (Fig. 57); small, cylindrical beetles (usually less than 1/3 inch in length) .... bark beetles: Scolytidae	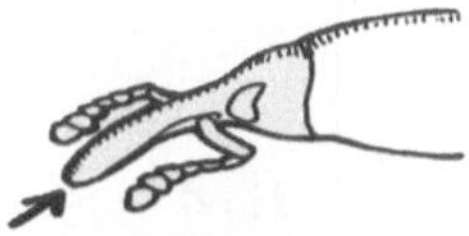 Fig. 56 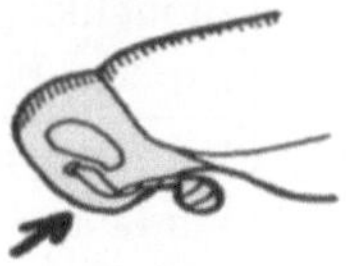 Fig. 57

Key to Major Families of Order Lepidoptera

S.No.	Characters of couplet	Figure
1.	Antennae threadlike and knobbed at tip (Fig. 58); no frenulum (spines on leading edge of hind wing which unite fore and hindwings in flight); ocelli absent(butterflies)...2 - Antennae of various forms, but usually not knobbed (Fig. 59); if antennae clubbed, then frenulum present (Fig. 60); ocelli often present (moths) 3	 Fig. 58 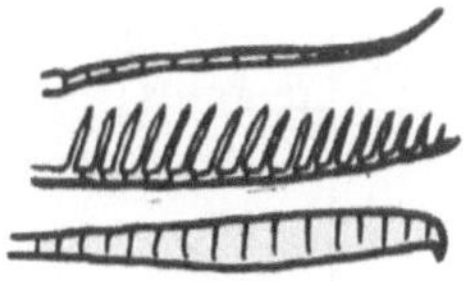 Fig. 59 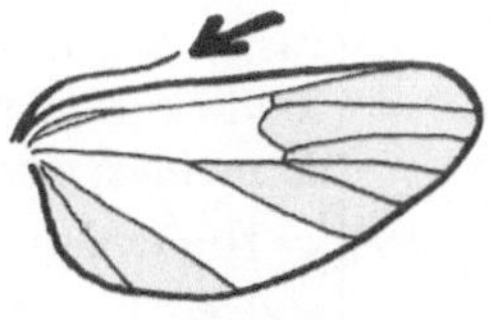 Fig. 60

2.	Small to medium sized butterflies; white, yellow or orange wings often marked with black.............................white butterflies: Pieridae - Small butterflies; metallic blue, green, copper or bronze colour markings and or tiny tail like projections (Fig. 61).. ....blues, copper and hairstreaks: Lycaenidae	 Fig. 61
3.	Hindwings with soft scales and small anal regions; palpi not unusually large and forming a snout like projection.. 4 - Hind wings with firm, fine scales and ample anal regions (Fig. 62); palpi often large and forming snout like projection (Fig. 63).............................. ....Flour moths, grass moths, etc.: Pyralidae	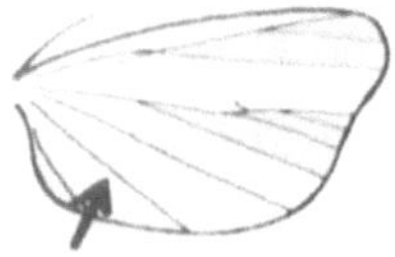 Fig. 62  Fig. 63
4.	Forewings usually long narrow, and pointed; hindwings usually short; body stout and tapered at both ends........................sphinx moth: Sphingidae - Forewings not long, narrow, and pointed; hind wings often nearly as large as forewings; body not usually stout & tapered...............................5	
5.	Labial palpi with third segment long and slender, usually tapering, the palpi upturned (Fig. 64).....................Gelachiid moth: Gelechidae - Labial palpi without third segment long and slender, usually tapering, or the palpi upturned...6	 Fig. 64
6.	A large part of the wings, especially the hindwings, devoid of scales except on veins and margins (Fig. 65).........Clearwing moths: Sesiidae - Wings normally scaled throughout and without extensive transparent areas................................7	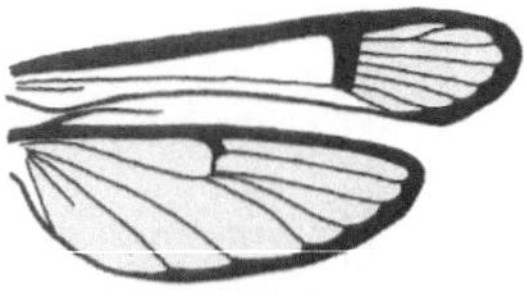 Fig. 65
7.	Hindwings with second and third vein usually fused basally to form single vein (Fig. 66); hindwings approximately equal in expanse to forewings; small to medium sized moths............8 - Hindwings without second or third vein stalked; hindwings usually distinctly smaller than forewings; small to large moths..........................9	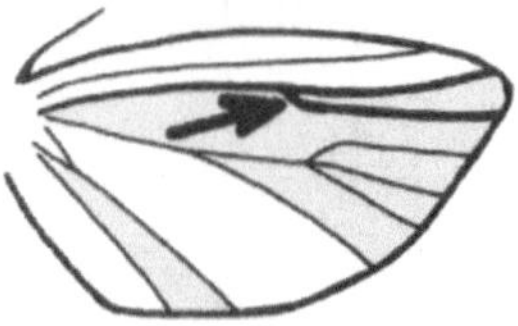 Fig. 66

8.	Hindwings usually with fringe of long hairs on upper side of basal part of vein 6 (Fig. 67); if fringe lacking, then veins 7 and 8 in front wing close together at tip (Fig. 68); outline of moth not typically bell shaped at rest.......................................fruit moths: Olethreutidae	 Fig. 67 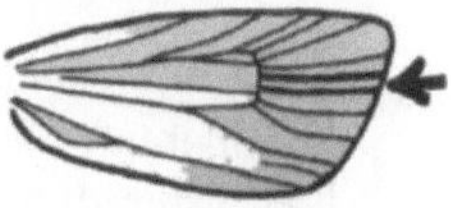 Fig. 68
	- Hindwings usually without fringe of long hairs on vein 6; if fringe present, then vein 5 and 6 in front wing stalked (Fig. 69) outline of moth typically bell shaped at rest (Fig. 70)...Leaf roller moths: Tortricidae	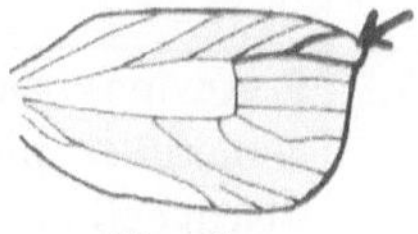 Fig. 69  Fig. 70
9.	Very small moths with narrow pointed wings; hind margins of wings with wide fringes of scales....10	
	- Medium to large sized moths without narrow, pointed wings; hind margins of wings without fringes of scales...11	
10.	Forewing usually without accessory cell; hindwing often has lump along leading edge near base (Fig. 71)..Leafblotch miner moths: Gracillariidae	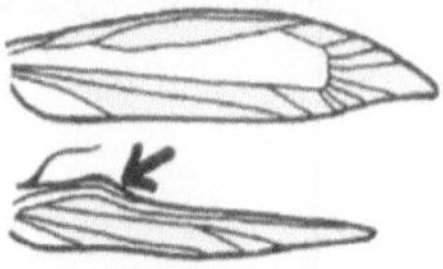 Fig. 71
	- Forewing usually with accessory cell (Fig. 72); hindwing tapers smoothly to apex..Clothes moths: Tineidae	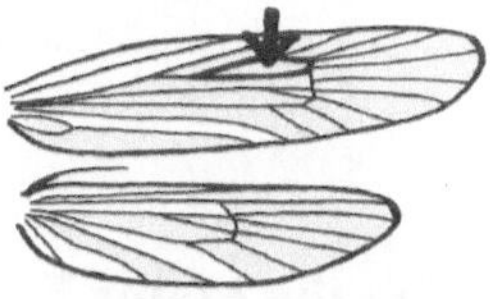 Fig. 72
11.	Proboscis rarely absent; body slender; legs slender, with few or no hairs; forewings marked with wavy parallel bands (Fig. 73)... measuring worm moths: Geometridae	 Fig. 73
	- Proboscis reduced or absent; body robust; legs well developed, either very hairy or spiny; forewings without wavy parallel bonds..............12	

12.	Ocelli absent; tympanum not developed on metepimeron (just below base of hind wing); frenulum absent; base of leading edge of hindwing not greatly expanded (Fig. 74)................................ ....Tent caterpillar moths: Lasiocampidae	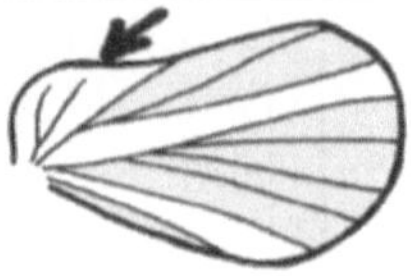 Fig. 74
	- Ocelli usually present (except Lymantriidae); tympanum developed on metepimeron (Fig. 75); frenulum present; base of leading edge of hindwing not greatly expanded..........................13	 WINGS Fig. 75
13.	Forewings brightly marked in contrasting colours, sometimes plain white or yellow; vein below discal cell in hindwing four branched (Fig. 76)..... ....tiger moths: Archtiidae	 Fig. 76
	- Forewings usually dull grey or brown; vein below discal cell in hind wing often appears three branched (Fig. 77)..14	 Fig. 77
14.	Antennae usually thread like; ocelli present; leading edges of wings quite straight (Fig. 78).. ....cutworm moths: Noctuidae	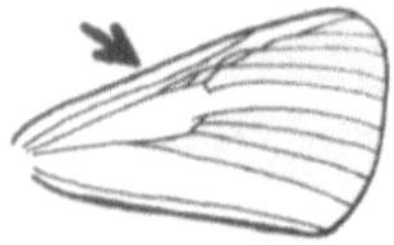 Fig. 78
	- Antennae feathery; ocelli absent; leading edges of wings rounded (Fig. 79)................................ ....tussock moth: Lymantriidae	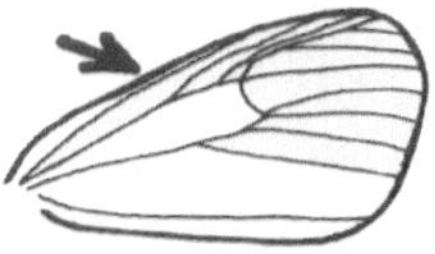 Fig. 79

Key to the Major Economic Families of Diptera

S.No.	Characters of couplet	Figure
1.	Antennae many segmented (long horned flies)........2 - Antennae with 5 or fewer segments, usually three (higher Diptera)...5	
2.	Wings with scales or dense hair on veins................3 - Wings without scales or dense hair on veins..........4	

3.	Wings very hairy giving an opaque moth like appearance (Fig. 80)............Sand flies: Psychodidae - Wings with scales along the veins (Fig. 81)........... ....mosquitoes: Culicidae	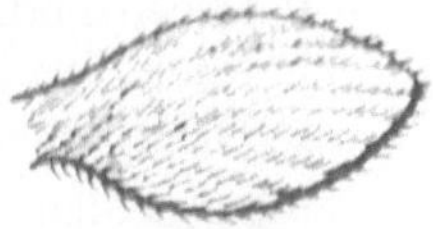 Fig. 80 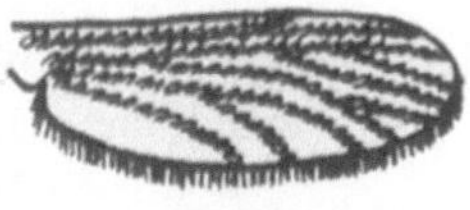 Fig. 81
4.	Tibiae with spurs; legs not unusually long and slender; costal vein not continuing around wing; piercing sucking mouth parts (Fig. 82).................. ....biting midges: Ceratopogonidae - Tibiae without spurs; legs unusually long and slender; costa continuing around wing (Fig. 83); without piercing and sucking mouthparts .... gall midges: Cecidomyiidae	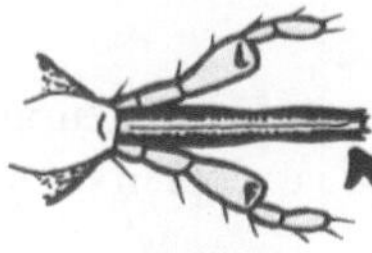 Fig. 82 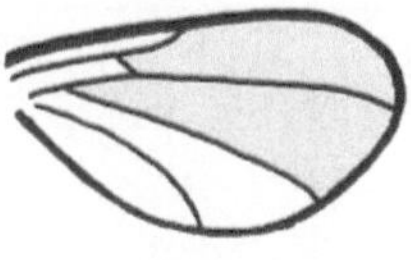 Fig. 83
5.	Head broad, thin, convex; antennae with third segment often toothed and distal portion ringed (Fig. 84).................................horse flies: Tabanidae - Head not broad, thin and convex; antennae without third segment toothed or distal portion annulate (rin ged)..6	 Fig. 84
6.	Wings with spurious (false) vein between third and fourth veins (Fig. 85).............flower flies: Syrphidae - Wings without spurious vein............................... 7	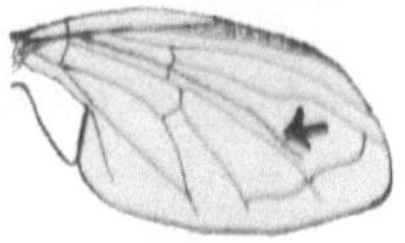 Fig. 85
7.	Second vein short, bends at right angles towards leading edge of wing; wings usually with bands of colour (Fig. 86)......................fruit flies: Tephritidae - Second vein not short and with right angle bend; wings not usually with bands of colour.................8	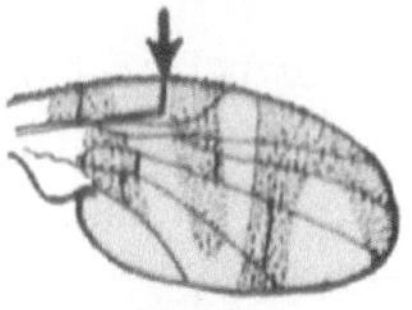 Fig. 86
8.	Mesonotum without bristles except above wings; characteristic weakened area across basal third of vein (Fig. 87)................................rust flies: Psilidae - Mesonotum with more or less complete bristle pattern. No weakened area across basal third of vein...9	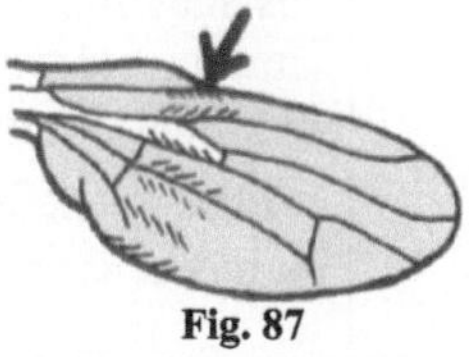 Fig. 87

9.	Postocellar bristles on top of head divergent (Fig. 88), if absent, the arista (large bristle in apical antennal segment) absent......Leafminer: Agromyzidae - Postocellar bristles not divergent; arista always present..10	 Fig. 88
10.	Hypopleural bristles absent (Fig. 89); arista plumose (with branches)..11 - Hypopleural bristles present (Fig. 90), arista bare except in some techinids.......................................12	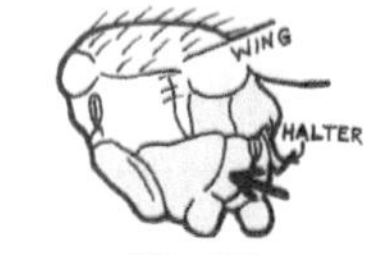 Fig. 89 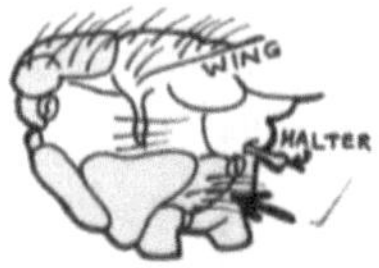 Fig. 90
11.	Sixth vein reaching wing margin, at least as a fold (Fig. 91)................rootmaggot flies: Anthomyiidae - Sixth vein never reaching wing margin, (Fig. 92)... .... House flies: Muscidae	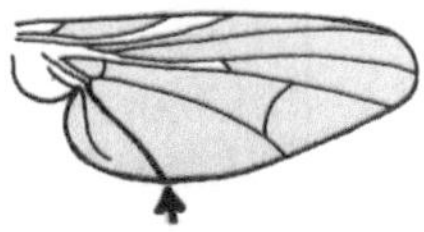 Fig. 91 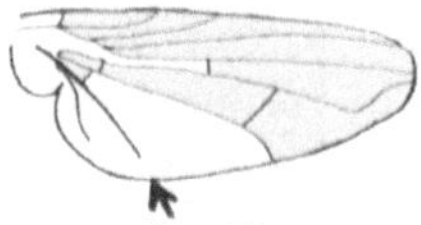 Fig. 92
12.	Mouthparts vestigial...13 - Mouthparts normally developed.......................14	
13.	Third and fourth veins diverge distally (Fig. 93); light yellow brown in colour .... horse bots: Gasterophilidae - Third and fourth veins do not diverge distally (Fig. 94); dark brown in colour .... bot and warble flies: Oestridae	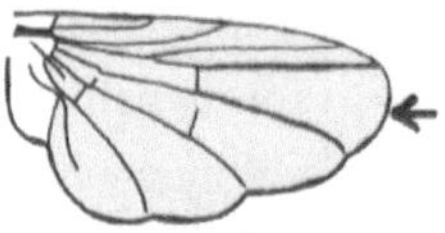 Fig. 93 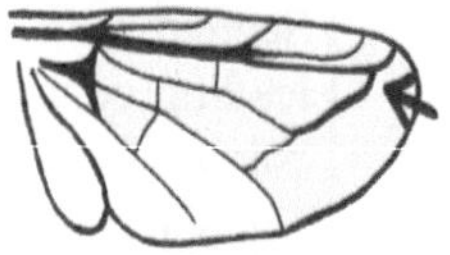 Fig. 94
14.	Head not fitting into emargination of thorax; winged Postscutellum, area of thorax immediately below scutellum developed (Fig. 95); arista usually bare, abdomen bristly .... tachina flies: Tachinidae - Post scutellum not developed; arista usually plumose, abdomen not bristly 15	 Fig. 95

15.	Thorax typically gray with three black lines, abdomen typically gray, never metallic .... Flesh flies: Sarcophagidae - Thorax not gray without black lines, abdomen with few exceptions, strongly metallic blue or green .... blow flies: Calliphoridae	

Key to the Major Economic Families of Hymenoptera

S.No.	Characters of couplet	Figure
1.	Abdomen broadly joined to thorax (non - constricted abdomen) .... 2 - Abdomen joined to thorax by a slender petiole or waist (constricted abdomen) 3	
2.	Body robust; abdomen not compressed laterally; fore tibiae with two apical spurs .... sawflies: Tenthridinidae - Body slender; abdomen compressed laterally; fore tibiae with one apical spur .... stem flies: Cephidae	
3.	Posterior trochanter consisting of two segments (Fig. 96) .... 4 - Posterior trochanter consisting of single segment...6	 Fig. 96
4.	Forewings with a conspicuous, dark, thickened spot (stigma) about midway on leading edge, venation not highly reduced .... 5 - Forewings without a conspicuous dark spot on leading edge; venation highly reduced (Fig. 97) .... chalcids: Chalcidoidea	 Fig. 97
5.	Two recurrent veins in forewing; small sub marginal cell (areolet) often present (Fig. 98) .... ichneumons: Ichneumonidae - One recurrent vein in forewing; small sub marginal vein not present (Fig. 99) ...... Braconids: Braconidae	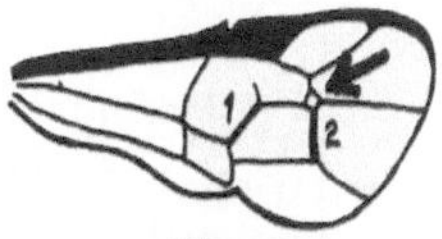 Fig. 98 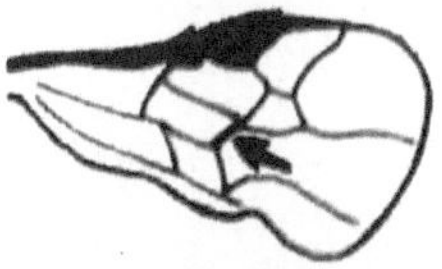 Fig. 99
6.	Antennae geniculate (elbowed); first 1-2 abdominal segments often with dorsal hump (Fig. 100) .... ants: Formicidal - Antennae not geniculate; first 1-2 abdominal segments without hump 7	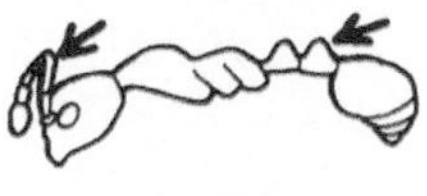 Fig. 100

7.	First discoidal cell very long (Fig. 101); wings usually folded lengthwise at rest .... social wasp, potter wasp: Vespidae - First discoidal cell usually not long; wings not folded length wise at rest 8	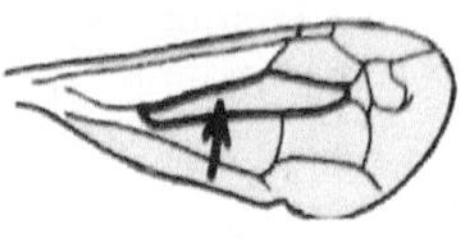 Fig. 101
8.	Body hairs unbranched; abdomen often petiolate (stalked); posterior angle of pronotum lobe like (Fig. 102) mud daubers, thread waisted wasps: Sphecidae - Body hairs on thorax branched (Fig. 103); abdomen not petiolate; posterior angle of pronotum not lobe like (Bees) 9	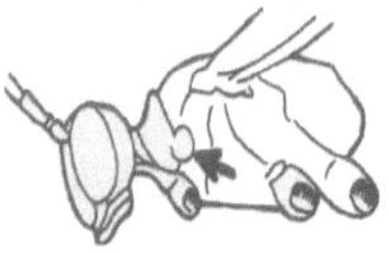 Fig. 102 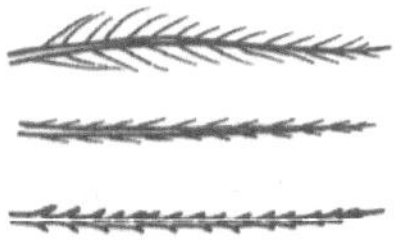 Fig. 103
9.	Basal vein strongly arched (Fig. 104); often with metallic coloures sweat bees, mining bees: Halictidae - Basal vein not strongly arched; without metallic coloures 10	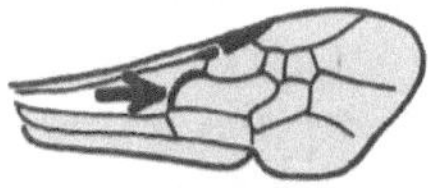 Fig. 104
10.	Two submarginal cells in forewing; abdomen typically boat shaped (Fig. 105) .... Leaf cutting bees: Megachillidae - Forewings with three sub marginal cells in forewing (Fig. 106): abdomen not typically boat shaped ..... honey bees, bumble bees, robust mining bees: Apidae	 Fig. 105 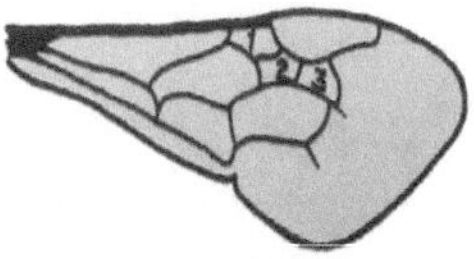 Fig. 106

15

Ethics in Taxonomy

Code of Ethics was observed by International Commission on Zoological Nomenclature for renaming of homonyms. The commission had issues declarations or opinions from time to time, which together constitute the start of a code of ethics in the field of nomenclature.

The first Declaration of the international Commission was as follows:

- When it is noticed by any zoologist that the generic name or specific name published by any living author as new is in reality a homonym, and therefore unavailable under articles 34 and 36 of the Rules of Nomenclature, the action is for said person to notify the said author of the facts, and give said author an opportunity to propose a substitute name.
- When the author of a newly discovered generic or specific homonym is dead, the discoverer is free to rename the genus or species as he likes. It is common practice to rename the category after the author of the homonym. However, this practice was never universal and it is not a matter of ethics.

An ethical problem arises when a taxonomist discovers a homonym outside the group he is working. Under such condition, it is the ethical procedure to permit the change to be made by someone who is familiar with the group and can judge if the change is required on zoological and nomenclatural grounds.

In cases of breach of ethics, Commission in its Lisbon meeting (1935), reaffirmed the Code of Ethics. But it was only at Paris (1948) that Commission added a recommendation to the rules. It stated that- "…. Condemning the selection of generic name of a word which purported to be an arbitrary combination of letters, but which when pronounced appeared to be a word or words in some language other than Latin, especially where those words had a bizarre, comic or otherwise objectionable meaning".

In contrast to the above recommendations, it was ruled that names "which can reasonably be regarded, in any language to give offence on political, religious or personal grounds" are prohibited.

The declaration 4, by the Commission in Monaco (1913) stated "the need for avoiding intemperate language in discussions on zoological nomenclature".

In the field of systematic zoology, a certain body of ethics has been built up as a moral sense of responsibility, courtesy and sensitivity to his fellow workers. The following points should be considered in ethics in taxonomy:

i. **Credit:** It involves giving proper credit or acknowledgement of all unpublished observations, determinations, photographs, drawings and data derived from others in a dignified manner. Previously published data should never be utilized in such a way as to appear original.

ii. **Collections:** If a collection by a taxonomist, becomes the basis of a published scientific study, it is not considered private collection. This is especially true if it contains a type material.

iii. **Borrowed material:** If a material is borrowed, the borrower is under ethical obligation, to study the material and return in good condition within a reasonable length of time. In certain groups, specialists expect the privilege of keeping a specimen in a series of many specimens.

iv. **Exchange of material:** The exchange of specimens is a simplest and least expensive method of building up a representative collection in any group. Exchanges are important in building up of complete synoptic series of genus and species.

v. **Relations with co-workers:** The worker has an obligation to his senior to maintain a relationship. This will permit free exchange of ideas and scientific information. This permits the worker to supplement each other's efforts, and to collaborate with them.

vi. **Suppression of data:** This is the responsibility of the author to mention those specimens which do not fit into the description or do not run out properly in the key, since these may later provide a most valuable clue in clarifying a taxonomic problem.

vii. **Undesirable features of taxonomic papers:** The author should avoid: (1) emotional phrascology (2) controversy (3) personal attacks (4) too much use of first person and (5) evaluation of his own work.

   Criticism is often exercised in the scientific method but it should be conducted in a dignified and constructive manner.

viii. **Letter writing:** The taxonomists may find it necessary to write letters requesting information, especially with regard to types. Such requests should be definite and specific asking for reprints. Requests asking for the specimens should also be specific accompanied by a statement of reason for the request. All letters should be carefully and tactfully prepared.

Although ethics in taxonomy are not part of the science but they are an important part of the relationship of the taxonomist with his fellow workers. This will influence to contribute his share to the advancement of entomology.

16

Hierarchy in Classification

Classification involves grouping organisms into a series of hierarchical categories. Actual method of establishing a classification consists in defining groups or categories on a hierarchical scale. The animals can be classified in a taxonomic hierarchy consisting of a series of categories of ascending ranks from species to kingdom. Each of these categories includes one or more groups from the next lower level. Therefore, all animals can be classified in a taxonomic hierarchy consisting a series of categories in ascending order from species to kingdom such as species, genus, family, order, class, phylum and kingdom.

According to the history of biological classification, **Aristotle,** a Greek philosopher classified different animals based on the habitat, characteristics etc. Later Swedish botanist **Carolus Linnaeus** introduced taxonomic hierarchy categories during 18th century and this system is followed globally.

Taxonomic hierarchy refers to the sequence of categories in increasing or decreasing order. Kingdom is the highest rank and species is the lowest rank in hierarchy. Taxonomic hierarchy is the process of arranging various organisms into successive levels of biological classification. Each of the level of the hierarchy is called taxonomic category or rank.

The function of taxonomic categories is to present diversity in nature in a concise manner. Every classification involves two steps-

a) Arranging lower units into groups.

b) Joining these groups in an ascending hierarchy of more and more unlike groups.

Taxonomic classification existed before the invent of theory of evolution and even now it is pursued without considering phylogeny.

Linnaenian Hierarchy

At the time of **Linnaeus,** only five hierarchic levels were recognized within the animal kingdom. These were :

i. *Classis*

ii. *Ordo*

iii. *Genus*

iv. *Species*

v. *Variety*

Later, two categories were added – *family* by **Butschli** in 1790 and *phylum* by **Hackel** in 1886.

Term variety was discarded in the taxonomic hierarchy of animals. As the knowledge of animals grew, it was necessary to make finer divisions. Mostly seven categories are in use viz. in case of honey bee:

Kingdom - Animalia
Phylum - Arthropoda
Class - Insecta
Order - Hymenoptera
Family - Apidae
Genus - *Apis*
Species - *Mellifera*

With time, two more categories have been incorporated, such as *tribe* between genus and family; and *cohort*, used by **Simpson** (1945), between order and class. The increase in number of known species together with the degrees of relationship between them influenced taxonomists to assign precise taxonomic position to species. This resulted in splitting of original basic categories and by adding additional categories between the seven basic one. Most of these are formed by combining the original names with the prefix super or sub. Thus there are super order, suborder, superclass and subclass etc. many other names have been proposed for higher categories, but none of them is in general use except the tribe between genus and family. Therefore, there are as many as 33 categories indicating hierarchy in classification. The most accepted categories are as follows:

1	Kingdom*	2	Subkingdom
3	Infrakingdom	4	Super phylum
5	Phylum*	6	Subphylum*
7	Infraclass	8	Superclass*
9	Class*	10	Subclass*
11	Infraclass	12	Supercohort
13	Cohort*	14	Subcohort
15	Infracohort	16	Super order*
17	Order*	18	Suborder*
19	Infraorder	20	Superfamily* (-oidea)
21	Family* (-idae)	22	Subfamily* (-inae)
23	Infrafamily	24	Supertribe

25	Tribe* (-ini)	26	Subtribe (ina)
27	Infratribe	28	Supergenus
29	Genus*	30	Subgenus*
31	Superspecies	32	Species*
33	Subspecies*		

Of these, only 18 (marked with asterisk) are generally followed. Names of tribes, subfamilies have standard endings which are added to the name. No standardized ending exists for categories above the family.

17

Palaeontology, Fossil Insects

Palaeontology is the science that deals with the life past geological periods. It is just the opposite of neontology which deals with the systematics of recent organisms. Fossils are the remains or parts of animal or plants that lived thousands of years ago which have turned into rock. They are preserved traces of living organisms in the earth crust.

Fossil History of Insects

Occurrence of insects in fossilised state provides a clear idea of the main order of insects. The oldest known insects' fossil appears to be of Apterygota and blattids seem to be the oldest living insects, which have almost lived without undergoing large scale evolutionary changes from the Carboniferous to the present day.

Some of the earliest apterygotes might have developed from the Permian or lower Carboniferous. The richest collection of fossil insects is from the vegetable deposits like coal, lignite peats and in amber entangled in the raisin. Most of the insects are known only through their wings, resulting in the study of venation. However, the amber fossil of Miocene presents a clearer picture. The best known epoch is undoubtedly the Permian while Triassic provides richest discoveries of the insect fossils.

In the middle of Devonian, an extremely small record of insect was found from the fossil peat bag of Rhynie in Scotland and identified as *Rhyniella precursor,* now regarded as similar to living collembolans with traces of ventral tube and retinaculum. The antenna with only three segments like scape, pedicel and flagellum. Since living Collembola have four segmented antennae, it may be assumed that they are derived from the primitive *Rhyniella.* Protura and Thysanura are not known from the earlier geological eras and are recognisable from the lower Oligocene.

The other insect fossils from the period of Carboniferous provide excellent example of insects of great size, with wings held horizontally while at rest. The principal fauna included Eodictyoptera, Protothoreptera, Protodonata, Megasecoptera and Blattids which still survive. Tyoptera are the most heterogenous occurring in France. They are characterised by small rounded head, setaceous antenna, prothorax with lateral notal expansions resembling fixed wing pads. Meso and meta thorax are sub equal, and the wings from these segments arise from broad bases and are elongate. The hindwings lack the fan-like anal lobe. The venation appears complete, with highly branched principal veins and irregular network of fine veins, the archedictyon. Abdomen 10 or 11 segmented, lateral lobes similar to those of prothorax and a pair of slender and elongated cerci and a median caudal filament. Example- *Stenodictya labata, Homaloneura bonnieri, Homoioptera woodwardi* and *Lithomantis carbonaria.*

Protothoreptera

Protothoreptera are early offshoot of the Palaeodictyoptera which include orthopteroid forms but lacking the prothoracic notal expansions, a more specialised prothorax and distinct fan like anal area in the hind wings. They have wings with very characteristic network of cross veins arranged at right angles to the main veins and a single anal vein. Protodonta unlike the living Odonata lack the true nodus and pterostigma on the wings. Some of the typical representation of wing venation in most of the Palaeodictyoptera with an archedictyon type are *Dieconeura arcuate* and *Protophasma dumasi, Dictyomylacris* and *Aphthoroblattina.*

The Protothoreptera from the lower Permian were initially believed to be ancestral to the Hymenoptera and both this and Protohemiptera cannot be considered as anything more than convergence. The protohymenoptera are reduced in size, slender and have delicate bodies. The presence of glossy, delicate wings with pterostigma and similarly in the disposition of the veins, sub-costa and radius are Hymenopteran features. Bothe the wings are closely alike in size, shape and venation. The Paramecoptera related to Mecoptera are represented by *Belmontia* and *Parabolmentia*, ancestral to Lepidoptera and Trichoptera. *Protoperlaria* appear ancestral to the living stone flies, Plecoptera. Protocoleoptera represented by *Protocoleus mitchelli* are both from the upper Permian beds, the later having a flattened elytra with straight sutural margins. However, the venational patterns of these early coleopteroid fossils is totally different from existing Coleoptera.

Megasecoptera

The Megasecoptera and Protodonta look closely related but the former with some elongated cerci, narrowed wing bases and wing without dense network of veinlets. They include species such as *Brodia priscotincta* and *Mischoptera woodwardi.*

Earliest Known Insects

As suggested by Ross (1963) two lines of descent, one line with pouched head, gave rise to Collembola, Protura and Diplura and the other lineage gave rise to Microcoryphia, Thysanura (Apterygota) and other Pterygotes.

Menton (1970) believed Diplura, Protura, Collembola and Pterygota as separate classes. She pointed out that Apterygotes are very different from one another in many respects. Diplura, Protura and Collembola have entognathous mouth parts while Thysanura and Pterygota possess ectognathous mouth parts.

Earliest known insect is a Collembollan *Rhyniella precursor* discovered by Hirst and Maulik in 1926 from the middle Devonian of Scotland. Most of the insect fossils are from Russia, Europe and North America, while some from southern hemisphere.

Oldest Fossil of Pterygotes

Eopterum and *Eopteridium* from upper Devonian were earlier considered as the oldest fossils among the apterygotes. Recently according to Rohdendrof (1972), they belonged to Crustacea. Therefore, probably Erasipteron (Odonata) and Poroprosbole (Hymenoptera) and Extinct Order Paleodactyoptera found in upper Carboniferous in the of middle Devonian of Scotland are the oldest.

Process of Fossilisation

Fossils are formed by a process called petrification. In this process hard parts or sometimes soft tissues of the body have been replaced by minerals. Iron, pyrites, silica and calcium carbonate are some common petrifying minerals. However, some animals are preserved without petrification in frozen state. Some small insects get embedded hard in the stone. Some fossils are found in the form of molds and casts. Both of these are similar to petrified fossils but are produced in a different manner. Mold fossils are formed by the hardening of material surrounding the buried organism followed by the decay and removal of body of the organism. The mold when filled by minerals and become hard are called cast fossils. These are exact replica of the original structure. Immediate burial is most important condition for the process of fossilisation.

Age Determination of Fossil

The age of fossil is determined by two methods:

a) Radioactive carbon dating.

b) Radioactive- potassium dating method.

Radioactive Carbon Dating

The method is based on the fact that in animals the ratio of radioactive carbon C^{14} and C^{12} remain constant during life. After death, the fossilisation ratio decreases as the C^{14} undergoes decay. By comparing this ratio in fossils the age of the fossil is determined. The half- life of carbon is 5568 + 30 years.

Radioactive- Potassium Dating Method

The transformation of radioactive-potassium K^{40} to organs is used for determination of age of the fossils more accurately as the radioactive carbon dating method can determine the age up to 25,000 years or about half-lives. The half-life of K^{40} is 1.3 billion years so that not only the age of the oldest fossil but the age of rocks can also be determined by this method.

18

Taxonomic Publications

The study of science is incomplete, if it is not made available to others. The most popular way to make the work wide circulated is through publication. It is most important for taxonomists that what is most needed in the world about their work, is to be published in time. In taxonomy, the publications may be in the form of book, pamphlets, journals, articles, scientific papers, symposium chapters, catalogues and check-lists etc. A taxonomist should publish his work in appropriate journals like population studies or biological studies etc. It should be mentioned in the title if the new research is published. Sometimes, new taxa of one country are included in the faunal work of another country without mention in the title about the country. This is bound to remain out of knowledge for future works. The major types of taxonomic publications are as follows:

Publication of New Taxa

These are ordinary descriptive papers of new taxa such as subspecies and genera. Such descriptions do not involve careful comparison of all related species. The published descriptions may be justified only when the names of the taxa are required for biological or economic works. These are also helpful in revision of any work where the new species can be easily fitted in the classification. In addition to this, short communications concerning with new records of various taxa are also quite useful.

Synopsis and Reviews

These include brief summaries of current taxonomic knowledge of a group. These should not include new taxa. Such publications actually bring together all the scattered information at one place and thus very useful for future revisionary or monographic works.

Examples

1. Srivastava, R.P., Bhatnagar, S.P and Gupta, R.S. 1976. Role of Taxonomy in Insect Pest Management. (Ed.) S.P. Bhatnagar, All India Symposium on modern concepts in Plant Protection. 1-8.

2. Nene, Y.L. 1988. Multiple Disease Resistance in Grain Legumes. Annual Review of Phytopathology. 26: 203-217.
3. Wilson, E.O. 1998. Biodiversity. National Academy Press, Washington, DC.

Taxonomic Revisions

These summarise and evaluate previous taxonomic works of a group incorporating new information. Some may follow monographic approach but not completely. Some are concerned with the arrangement of previously published taxa. They may be at species or generic level.

Examples

1. Boucek, Z. 1988. Australasian Chalcidoidea (Hymenoptera) – A Bio-systematic revision of genera of fourteen families, with a reclassification of species. CAB International, Wallingford, U.K., Cambrian News Ltd. Aberystwyth, Wales, pp130.
2. Capek, M. 1970. A new classification of the Braconidae (Hymenoptera) based on the cephalic structures of the final instar larva and biological evidence. The Canadian Entomologist 102: 846- 875.

Monographs

These are the publications involving full systematic description of all species, subspecies, and other taxonomic units with thorough knowledge of the species, subspecies and immature stages in the group and comparative anatomy, metamorphosis etc. of the group. It contains exhaustive bibliography as it assembles all such work.

Example

1. Agrawal, Neerja. 2020. Horticultural crops and Fruit flies of India. Pub. Kalyani Publishers, New Delhi, pp1-173.
2. Narayanan, E.S. and Batra, H.N. 1960. Fruit flies and their control. Pub. ICAR, New Delhi, pp 10- 65.
3. Noyes, J.S. 1985. Chalcidoids and biological control. Chalcid Forum, 5: 5-7.

Faunal Studies

Faunal studies are detailed description of fauna of a single region. The study is limited to a single group of animals of an area or any major group.

Example

1. Bey-Bienko, G.V. 1936. Insects Dermapteras, Fauna de 1, URSS, Moscow and Leningrad, 10: 1-212.

2. Fauna of British India (Now Fauna of India). Tayler and Francis, London. These run in many volumes in many group of animals. Published since 1888.
3. Fauna de France, 1921-1967. Vol. 1-68. Office Central de Faunistique, Paris.

Atlases

They depict comparative characters of animals in picture form. The purpose of atlases is purely taxonomic and includes semi diagrammatic drawings, full, halftones or coloured plates.

Example

1. Ferris, G.R. 1937-1950 (5 volumes). Atlas of the scale insects of North America. Stanford University Press, Stanford University, California.
2. Ross, E.S. and Roberts, H.R. 1943. Mosquito atlas. American Entomological Society of Philadelphia, 1: 1-44, 2; 1-44.

Catalogue

A catalogue is mainly an index to taxa arrangement in such way as to give a vivid picture of references for both Zoological and nomenclatural purpose. According to Blackwelder (1969), a catalogue contains the information as given below:

1. The original description reference.
2. Later reference.
3. Synonyms with reference.
4. Range.
5. Type locality (also its repository).
6. Type of genus.
7. Type of species.

The taxa are arranged in chronological order. The catalogue gives authentic information and therefore, care is taken in its preparation. A person must have thorough knowledge of the methods and source of Bibliography.

Example

1. Krombein, K.V., Paul, D.H. Jr., Smith, D.R. and Burks, B.D. 1976. Catalogue of Hymenoptera in America North of Mexico. Smithsonian Institution Press. Vol.1-3.
2. Shenefelt, R.D. and Maish, P.M.1976. Hymenopteran Catalogues Part 13. Braconidae 9: 1263- 1424.

3. Townes, H., Townes, M. and Gupta, V.K. 1961. A catalogue and reclassification of Indo-Australian Ichneumonidae. Memoir American Entomological Institute., 1 & 2: 1-1210.

Checklist

These are abbreviated synopsis of groups of animals and present a list of names. Mostly a checklist is popular in terms of butterflies, birds and mammals.

Example:

1. McDunnough, J. 1938- 1939. Checklist of the Lepidoptera of Canada and The United States of America, Part I. (Macrolepidoptera). pp. 1-275; Part II (Microlepidoptera). pp 1-171.
2. Wood, A.M. 1989- 1992. Insects of economic importance: A checklist of preferred names. C.A.B. International, Wallingford, Oxon OX10 8DE, UK. pp 149.

Field guide

Field guides are prepared for the non- taxonomists, to help them to identify common animals in the fields. These can be easily understood by easy key characters. Some field guides are specially designed in the form of pamphlets crop wise, for periodical check by field workers for possible entry of new crop pests.

Example

1. Peterson Field Guide Series: Putnam's Nature Field Books and Jaques Pictured Key Nature. Pub. William C. Brown Company, London.
2. Menon, V. 2003. A field guide to Indian Mammals. Pub. Dorling Kindersley (India) Pvt. Limited.

Manual

These are meant for students or laymen, published in simple language including key characters for common species of animals. Sometimes they are published as a comprehensive work.

Example

Agrawal, Neerja. 2021. Fundamental Entomology: A Practical Manual. Jaya Publishing House, New Delhi. pp 1-146.

Handbooks

The handbooks are used either as field guides, manuals or occasionally comprehensive volumes on a group of animals of relatively comparative taxonomic treatment.

Example

1. Grainge, M. and Ahmad, S. 1988. Handbook of plants with pest control property. John Wiley, New York.
2. Ali, S. and Ripley, S.D. 2001. Handbook of birds of India and Pakistan together with those of Bangladesh, Nepal, Bhutan and Ceylon. Vol. 1-9. Oxford University Press, Delhi.
3. Sehwoerbel, J. 1987. Handbook of Limnology. Ellis Harwood Ltd., Chichester, UK.

Treatise

Some of the handbooks are known as Treatise.

Example

1. Grasse, Pierre- P. 1948-1979. Grasse's Traite de Zoologie. Vol. 1-52.
2. Moore, R.C. (ed.) 1953. Treatise on invertebrate Palaeontology, Geological Society of America, University of Kansas Press.

19

Hybridisation and Speciation

Hybridisation

Hybridisation occurs when two species cross to form a new species. If they are named before their hybrid nature has become apparent, the names would become invalid as soon as the hybridity of the specimen has been established.

Hybridism can be of three types:

1. Sympatric Hybridisation

When individuals in two regions overlap in an otherwise well-defined species, they are called sympatric hybridisation. This is common in bird of paradise and humming birds. This hybridisation is listed as cross of two parental species.

Example- *Tetrao urogallus* X *Lyrurus telrix.*

2. Allopatric Hybridisation

In this case, the two parental species are conspecific. If hybridisation is defined as crossing of unlike parents, it is difficult to make a sharp distinction between the hybridisation of subspecies and allopatric species. It is often convenient to consider as allopatric hybridisation if interbreeding between different species in different regions develops.

3. Amphiploidy

This mostly occurs in plants. Hybridism in plants may lead to production of polyploidy that combines the chromosome sets of two parental species. This may give rise to a new population which is reproductively isolated from the parents and which leads to formation of a new species provided it is sufficiently fertile and able to compete with other species including its parents.

Speciation

Darwin examined Walsh's concept, rejected it and defined the term species as one arbitrarily given for convenience to a group of similar individuals. The first edition of **Dobzhansky's** Genetics and Origin of Species (1937) greatly stimulated interest in the nature of species. The results have been a large number of contributions to the theory of speciation and an effort by taxonomists to apply this new thought in their practical work, especially in that pertaining to closely related forms.

Later, **Mayr, Stebbins, Emerson, Jepson, Linsley** and **Usinger** offered comprehensive treatments on the nature of species.

According to **Mayr** "speciation means the formation of species". Speciation in its restricted modern sense, however, means splitting of a single species into several that is, multiplication of species.

Types of Speciation

Ecological Speciation

Ecological speciation is referred to all those forms, that pertains to the relationship of a population to its environment.

Independent breeding population are species. Such populations of one region may resemble one another so closely that they are difficult to segregate and it is frequently difficult to relate properly more or less similar populations from different regions. In the present scenario, it is difficult to examine a new species.

Walsh (1863) was the first entomologist to define the term species more objectively than Darwin. He considered the taxonomy of closely related species as a special problem, four years after the publication of Darwin's origin of species. He explained that two supposed species are distinct as they do not mix sexually together if geographically separated. Some of the species barely differ morphologically.

Mayr, in his "Principles of Systematics" noted that all species criteria and all species are based on three basic concepts:

a) Typological species concept.

b) The non- dimensional species concept.

c) Polytypic or multi- dimensional species concept.

The common thing in above all these concepts is that the species is recognised as the common breeding unit and parthenogenetic species are fundamentally different from bisexual species.

a) Typological species Concept

When species status is determined solely by the degree of morphological difference, it is the typological species concept. In this concept, species could be defined statistically. Any taxonomist working in the absence of biological data or field experience is in this position. The concept can breakdown in instances of extreme sexual dimorphism but is the only one applicable to organism with sexual reproduction.

- **According to Darwin** "The term species was given to a set of individuals closely resembling each other". This concept results in the morphologically defined species.
- **Thorpe** defined "the species as a group of individuals distinguished from all other groups by commonly possessing certain structural characters". This concept was unsatisfactory because it leads to a purely arbitrary definition.
- **Simpson's** concept can be applied by the taxonomists to a limited set of collection, without the knowledge of the species in nature. But this concept also failed to deal with infra specific variations.

Therefore, the typological concept was rejected because the morphological criteria was more useful in taxonomy. The morphological characters can be evaluated to produce an objective taxonomy for closely related forms. The morphological characters that distinguish species are of the same order and other variations within the species. For example, the male genetalia are more useful than other body parts in separating insect species, though in many groups they offer no character for the separation of certain species.

b) The Non-Dimensional Species Concept

It is based on distinction and not on differences. The distinction must be of individual insect of conspecifics from other species and the manifestation of this in reproductive isolation between two populations from the taxonomic point of view.

This concept is based on the relationship to one another of coexisting natural population in a non-dimensional system i.e. at a single locality at one time. It is based not on differences between individual organisms but on the discreetness due to reproductive isolation of populations that are composed of freely interbreeding individuals. The criterion of reproductive isolation is absence of interbreeding between such populations. It is non- arbitrary and is of utmost importance to understand sibling species and polymorphism in general.

c) The Polytypic or Multidimensional Species Concept

This recognises that species are not restricted to a single point locality but may have extensive ranges across whole continents or archipelagos. Their biological characteristics may vary from place to place. It is often hard to apply the second concept in such situations and taxonomists may have to fall back on the morphological criteria of the first concept.

This concept considers the species as a group of population that are actually or potentially capable of interbreeding natural populations which are reproductively isolated from other such groups (Mayr). In laboratory, the breeds might hybridize successfully if isolated by ecological or ethological factor, and come together in nature.

The Isolating Factors are of Two Types

a) Extrinsic (Spatial or Geographical in parts).

b) Intrinsic (Reproductive or physiological).

Extrinsic Factors

Spatial isolation is that caused by natural barriers, the kind that has been overcome by the insects of foreign origin that are established in North America.

Intrinsic Factors

Intrinsic factors are ecological which prevent the mating of potential mates; the ecological factors cause males and females of different species to be unattractive to one another causing sexual isolation. Mechanical and physiological factors prevent copulation and fertilization. This causes varying degrees of unviability and sterility in hybrid offspring.

As the species very rarely cross-mate under natural condition, sexual and ecological isolation are the effective factors in insects.

Breakdown of intrinsic isolating mechanism results in gradual infiltration of the germ plasm of one species into that of another known as introgressive hybridization. It is very rare in insects but the phenomenon occurs in plants. For example, butterfly *Colias eurytheme* Boisduval has greatly extended its range in eastern United States where it hybridizes and competes with *Colias philodics* Latreille to the disadvantage of the latter.

Thorpe accepted the typological species concept. It follows that the sympatric physiological or biological races of the typological species concept are sibling species in the non- dimensional concept.

20

Taxonomic Collections Curation and Preservation

Insects are present everywhere. In the world where life is possible, insects can be found. To begin with, it is not necessary to go to distant places. Insects can be collected from garden, cultivated fields or even in our house. For the collection of insects, we must have some idea of the following:

1. How they live.
2. What they require at different stages of life.

A. Complete Metamorphosis

There are four stages

i. Egg stage
ii. Laval stage
iii. Pupal stage
iv. Adult

B. Incomplete Metamorphosis

There are three stages:

i. Egg
ii. Nymph
iii. Adult

Catching of adult stage is only one part of collection which give very little view of their life cycle. To know an insect, the young insects are kept alive and reared to maturity. Beside this, things that insect of all stages are concerned with are:

1. Light, 2. Warmth, 3. Food, 4. Moisture and 5. Shelter

Knowing habits of insects makes collection much simpler. Before starting collection, stand and watch the insects for a while. See how pollen loving insects sit on the flower. Some, like bees, go from flower to flowers without wasting little time. Others like, butterflies, some moths and many flies love the warmth of Sun and spend much of their time just basking, either quite still, or slowly opening and closing the wings. On the other hand, hovers are able to remain poised in the air, apparently motionless but the wings moving at very high rate.

The leaves and stems of the plant shelter many insects. Underneath the leaves clusters of eggs can be seen or hanging pupae as well as many insects. On foliage a number of adult insects may be seen some of which are carnivores and look for a prey.

Insects that fly can be easily caught by the net, but smaller insects that keep still, or hide away are more difficult to collect.

General Collection

For making general collection one should search the follows places:

1. **Scrub land:** With low bushes of different species, long and short grasses.
2. **Open hill side:** At this place the insects are not so crowded together as in scruby land. At hillside most of the collection is done on the tree trunks or on the ground at the foot of the trees. Bigger insects can be stalked and trapped. The new forest is a good locality and tremendous local variation in the number of insects can be seen.
3. **Grass Land:** It can be looked at following places

a) **Low Land Pasture:** It has a much varied insect fauna. The number of individual insects may be mainly butterflies, bee or beetles. Most of the insects are hidden in the grass or around its root. Sweeping is the best way to collect them.

b) **Open down land:** It has a fauna of its own. Flowers are usually plentiful, and there are more insects than on many lush pastures. Butterflies are the common insects at this place.

c) **Heath land:** It has fewer flowers but more scope for sand living, fossorial insects, especially Hymenoptera and beetles.

d) **Limestone fells:** They have very poor grass where the species of scale insects can be collected.

All these are more general way of collecting in daylight, but there are other places where a special and some peculiar fauna can be seen. All kind of rubbish and debris have their own insects as rotting and decaying are part of natural cycle. Insects play an important role in this process. These are:

1. **Saprophytic insect:** Animal dung attracts many insects to lay eggs and to use it as food for their larvae.

2. **Parasitic insects:** Parasitic insects can be caught on the wing while waiting near a bait animal, such as cow, dog, horse, monkey etc.
3. **Ectoparasite:** Insects that live on the outside of other animals. Fleas and lice are the best known example of this group.

Catching and Traping of Insects

After finding and observing the insects, next step is to obtain some of them for further study either dead or alive. For catching insects some equipments are required such as nets, aspirators and tubes etc. There are various types of nets used for collecting different insects, but the best collection is done by a general purpose net.

NETS

An insect may be captured by net in the following ways:

1. By catching it in flight.
2. By stalking the insects until it settles, and then dropping the mouth of the net over it.
3. By sweeping, by swinging the mouth of net through grass or soft heritage, so that the insects are disturbed and are trapped in leg of the net.
4. By beating, that is holding the net beneath bushes and beating the foliage with a stick so that insects fall into the net.
5. Insects in the pond and streams can also be captured by net.

General Purpose Net

The general purpose net is either circular or pear shaped (Fig. 20.1). The frame should be 30.5 - 46 cm across. The pear shaped frame makes it easier to swing the net close to branches, tree trunk, wall or the ground. This kind of frame is also easier to put one's head into the net to inspect the catch. A frame that can be folded, or taken apart into several pieces, is convenient to carry in the pocket or under the arm when not in use.

A simple frame is made from three or more pieces of cane, joined together either with brass hinges, or with brass tips and sockets pushing into each other. Such a frame is light and strong. Metal frame are made by either a single ring of heavy metal rod or two or more sections joined together. The best all-round frame is made from two pieces of steel strip in round or pear-shaped design. When folded it makes a straight object, like an umbrella.

The bag should be at least twice as long as the diameter of the frame, so that with a twist of the wrist it can be closed over the frame. The material from which the bag is made should be light and soft made of mosquito netting.

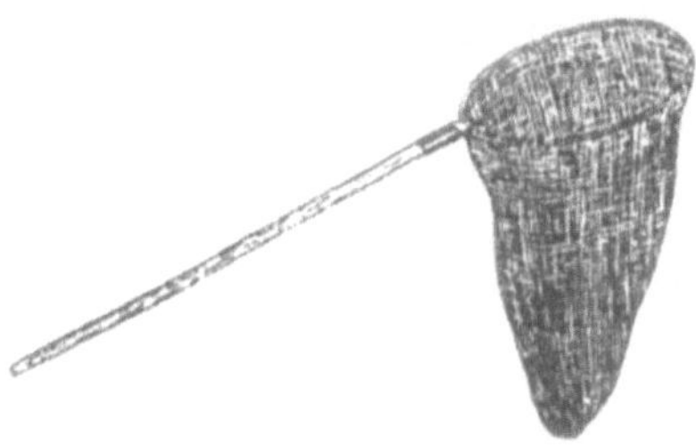

Fig. 20.1 General perpose net

For various type of insects' collection, if net is to be used, a general purpose net can be improved upon as follows:

1. **For flying insects only:** The aperture should be big, the net frame light in weight and bag of open mesh.
2. **For stalking individual insects:** A smaller frame is most advantageous because the ground is generally uneven, a light flexible frame is easier to press down the ground. Bigger the frame, it is more likely to leave crevices.
3. **Sweeping:** The frame should be big for bigger collection and also the collector can put his head and hands inside to inspect the catch. Both the bag and frame should be very strong. Bag should be of denser material. This frame can also be used on the water net with an inter changeable bag.
4. **Beating:** For this purpose, a simple beating tray can be made from canvas spread over strips of wood bamboo on the same line as making the kite. An old umbrella will also serve the purpose. Beating is intended for catching crawling insects.
5. **Water Collection:** A separate net for collecting in water is required. It should be stronger and heavier. A square frame is more advantageous.

Handles for Net

For air collection short handle of about 2 feet is ideal because it is useful in carrying from one place to other. Water nets need a long, strong handle to get enough reach.

Transferring The Insect from Net to Container

Water insects are helpless when the water is drained away and they can easily be tipped into jar. But other insects are more active and can be transferred in the following ways:

1. **Tubes:** Useful size of tubes are 10 cm x 2.5 cm or 5 cm x 1.25 cm with cork to fit. Resting insects, plant bugs, mosquitoes can easily be caught by putting a tube directly over them.

2. **Collecting bottles:** It is a simplest variation of the tube. In this case both ends of the tube are open which is passed through the cork of an oval bottle which can easily be held by hand. The open end of the tube is placed over the specimen and when the insect has walked in the tube, the end is corked. This bottle can be converted into a killing bottle also (Fig. 20.2).

Fig. 20.2 Collecting bottle

3. **Aspirators:** After the simple tube and bottles, these are the most useful of collecting equipment. These are used to collect small insects from the net or directly from foliage or even from the ground. A suction bottle is made by taking a small bottle, like collecting bottle, but fitting of the cork with two small tubes. The two tubes are bent at right angle or to any other angle as per convenience of collector. By sucking at the rubber tube, small insect can be drawn into the bottle (Fig. 20.3).

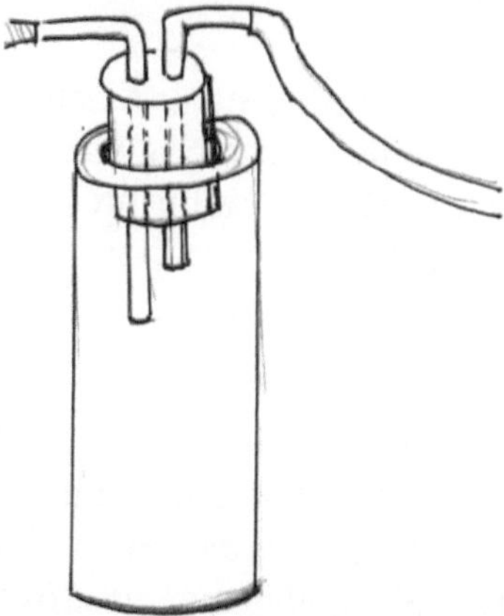

Fig. 20.3 Aspirator

Collecting Insects from Light Trap

Most of nocturnal insects are attracted to light. These may be moths, many kind of midges, some beetles and lacewings and stone flies. Collection on light trap is affected by a number of factors and weather conditions i.e. moonlight, temperature, humidity, wind velocity etc.

The simplest form of light trap is the box trap which can easily be made. Five of the six faces of the box are solid. The sixth is blocked by two over lapping sheets of glass which slope inwards. Thus an insect seeking the light is guided through the narrow slit between the sheets of glass, but it has a small chance of finding the slit in the opposite direction. An open killing bottle and lamp may be placed inside the box so that the fumes coming from bottle may kill the insect (Fig. 20.4). A mercury vapour lamp is more useful because it gives the light with high content of ultra violet.

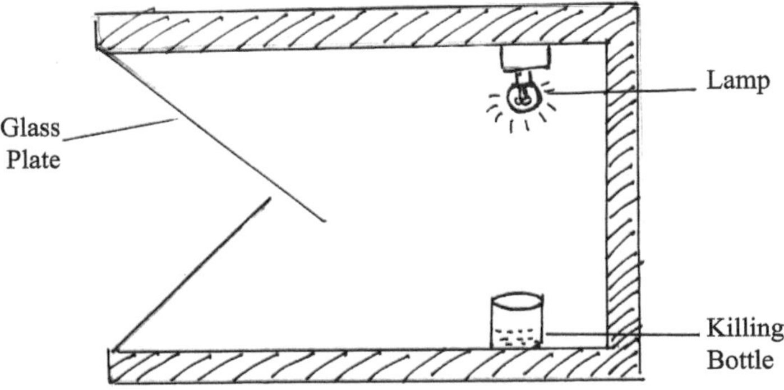

Fig. 20.4 Light trap

Other Types of Trap

1. **Baits and Bait trap:** For this purpose, baits for insects, either a natural substance that is known to attract them or, a synthetic substance giving off the same odour in even greater concentration is used. The bait is prepared with a fermenting mixture of sugar, molasses and beer.
2. **Rot-holes in the tree:** Different types of insects and their immature stage can be collected from these places.
3. **Pit falls:** Crawling and running insects may be caught in baited traps sunk into the ground so as to form pitfalls from where insect cannot climb out again, because the sides are too steep and smooth. Glass jars can be used to make the most convenient containers. A piece of wood is supported over the mouth so that frogs do not get in and eat the insects (Fig. 20.5).

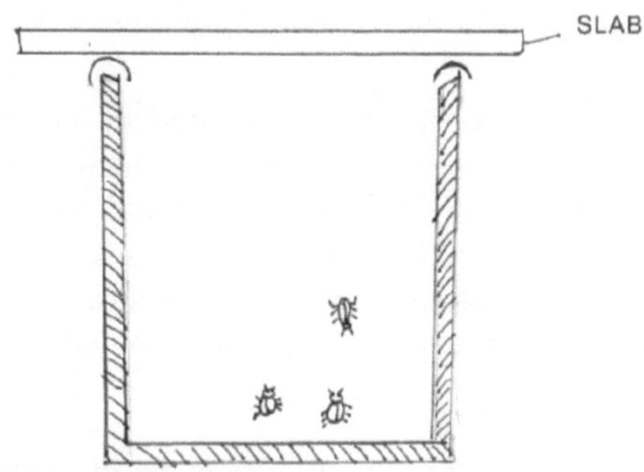

Fig. 20.5 Pit fall trap

Rearing of Insects

For the collection of immature stages, it is necessary that insects should be reared in laboratory. The insect that can be successfully reared continuously in the laboratory are those whose immature stages are all passed in a uniform and stable medium. This include insects living in decaying or fermenting materials (fruit flies), dry stored products (beetles, moths) etc.

The insects can be reared in a glass jar covered with a muslin cloth or netting on desired humidity and temperature. It is better to put some sand or soil at the bottom with one or two twigs of plant so as to provide same natural condition. Moisture can be provided by damping the soil daily. A piece of sponge soaked with water or a small glass tube filled with water and its mouth plugged with cotton inside the cage is useful to maintain moisture.

Larval Rearing

For rearing in laboratory, the immature stages of insects such as nymphs, larvae or pupae are kept alive until they develop into adults. The larval stages can be reared on natural food or on artificial food. A number of artificial foods for rearing larvae of house flies may be used for other insects too. They are mostly made up from bran, oats or powdered milk which is moistened. Carnivorous larvae can be reared on ground meat, provided it does not contain too much fat which produces acids that are less favourable for larval growth. Larvae can also be reared on an entirely artificial medium consisting of Agar, baker's yeast and salt.

Pupation

Suitable site for pupation is very essential for successful rearing. Caterpillars which feed on leaves in the open need support for pupa and some form of shelter also. Fruit flies reared in bottles climb up and pupate on the glass above the culture medium. An artificial site can also be provided by putting a card board inside the bottle.

Many insets pupate in soil. For this purpose, sand may be used in the jar instead of soil. Insects that are usually resistant to drought in the pupal stage (eg. blow flies) will pupate successfully in any crumpled paper and a most useful material for this purpose is the rough cellular material that is used for packing eggs. It will also be useful for lepidopterous larvae which spin a silken cocoon. Sufficient moisture should also be provided for successful pupation.

Rearing of Leaf Mining Insects

In this case the infested leaves are put straight into an airtight tin or screw topped jar to avoid loss of moisture from the mine. As soon as the insect is seen to have pupated, it is removed from the leaves. The pupae are kept in small labeled tubes. When they emerge they are killed and mounted with the empty pupal skin on the same mount.

Sleeving of Larvae

When it is not possible to remove part of the food plant of a larvae and grow this in a small cage, the damaged twigs where larvae are present, may be confined in a sleeve of fine muslin by enclosing the twig. Larvae that do not pupate in the soil will make their cocoons inside the sleeve and the twig can be cut off and moved to a separates cage, ready for the emergence of adult insect. For those which pupate in soil, when the larvae are fully fed, one end of the sleeve is opened and tilted down the rim of a pot or jar containing soil. When the larvae move down in the jar, it is transferred to a cage for adult emergence (Fig. 20.6).

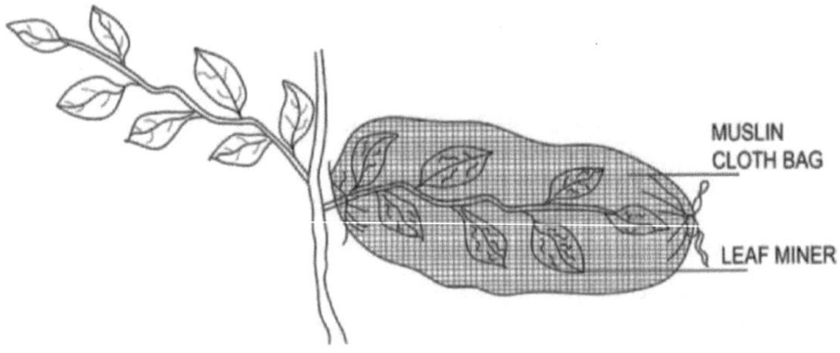

Fig. 20.6 Sleeving of larvae

Killing of Insects

Insects caught in various catches may be killed in killing bottle by various methods:

1. Cyanide Bottle

Killing bottles are made in a strong bottle or tube with a specially well fitted cork. In the bottom a layer of 5mm thick broken potassium cyanide is placed. Powdered plaster of Paris is used to fill the space between the lumps of cyanide. Some more plaster of Paris is mixed with enough water to make a slurry and this is poured over the previous mixture. The setting of plaster is an exothermic reaction in which much heat is generated and much water vapour is released. It is necessary, therefore, to set the bottle or tube aside, open for a day or two until the plaster is hard and dry. When the drying is finished, the cyanide will be held safely under a hard but porous layer of plaster and the gradual decomposition of the potassium cyanide will release hydrogen cyanide, which will percolate through the plaster and fill the interior of the bottle (Fig. 20.7). The cyanide bottle is liable to release vapour when in use, therefore, a circular piece of blotting paper may be placed on the top of the plaster. This should be renewed when damp. The used blotting paper must be burned.

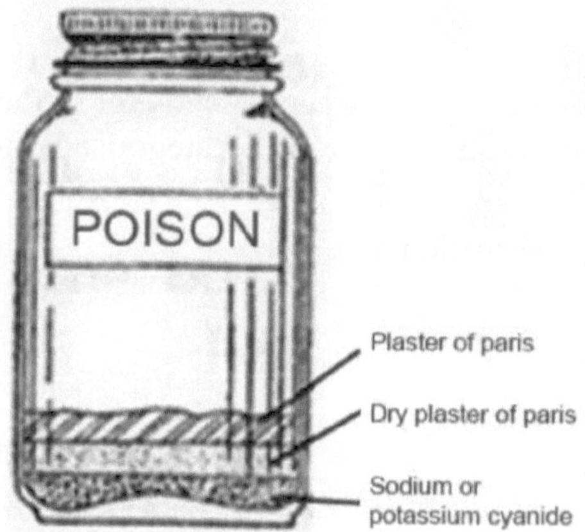

Fig. 20.7 Killing bottle

2. Liquid Killing Agents

Tubes or bottles are provided with a layer of plain plaster of Paris (without cyanide). This type of bottle is permanent and can be used over and over again with any of the liquid killing agents.

Before going out for collection, a little of the liquid is poured on the plaster, taking care that only limited amount of liquid is poured, that the block can soak up. If it is more, the specimen will get wet and spoiled. The cork is replaced tightly and bottle is ready for use. It is important that excess liquid will harm the person because if the bottle is left in sun or kept in warm pocket or held in hand for longer.

Killing Agents

1. Ethyl acetate ($CH_3COOC_2H_5$)
2. Ammonia (NH_4OH)
3. Benzene (C_6H_6)
4. Chloroform ($CHCl_3$)
5. Carbon tetrachloride (CCl_4)
6. Trichloro-ethylene (C_2HCl_3)

Destruction of Killing Bottles

It is important in case of cyanide bottles. The old cyanide bottles should be broken and buried in damp soil, taking care that dog or other animal may not dig it. Never get rid of it by throwing it into a stream.

Other kind of bottles with an absorbent layer for using liquid killing agents may be used over and over again.

Preservation of Insect

After killing the insects, it is necessary to preserve them permanently for further study or display. Methods of permanent preservation can be categorised into 4 categories.

1. Preserving the specimen dry (dry preservation).
2. Keeping it in liquid.
3. Immersing it in a resinous material.
4. Mounting on a microscope slide.

1. Dry Preservation

Insects with their external chitinised skeleton have great advantage over mammals or birds that most of them can be left out to dry naturally without offensive decay. The pinned and dried specimen is the most common and useful for ordinary purpose. Dried insects are extremely brittle and can be broken even with a slight touch, therefore, all manipulation of them should be completed while they are still fresh. That is why pinning as early as possible after collection is advised.

Drying and handling can be postponed by using ethyl acetate as a killing agent and having the specimen in the vapours. Simply keeping the specimens in a tin with moistened blotting paper will delay drying. Any fleshy specimens should be split open and the body contents should be removed, before decay can start and turn the specimen black.

When collecting the specimens daily for longer period, it is not possible to pin the specimen daily out of laboratory. After one or two days of collection the specimen will be quite hard and brittle. Such specimen requires relaxing or re-softening of the external skeleton for further studies.

Relaxing

This can be done simply by keeping the specimen in a humid atmosphere until the integument absorbs enough moisture again to become soft. For this purpose, relaxing tins are prepared which are made of zinc or other rust less metal and these have tightly fitting lids. In the bottom there is fairly thick layer of cork and other synthetic cellular material of an absorbent texture. The absorbent layer is moistened with plain water. An exposure of 12-24 hours will be sufficient to relax most specimens but exceptionally big insects, old ones or greasy ones may take longer. To prevent mold or fungus in the relaxing tin, it is desirable to put phenol or ethyl acetate.

Setting

This is also called stretching or spreading. This is the craft of arranging the soft, relaxed insects with its wings extended horizontally and allowing it to harden in that position.

Setting boards: These can be bought ready-made or easily made. The side boards are covered with cork with a groove in the middle to take the body and legs of the insect. This groove has underneath it a sheet of cork to allow an entomological pin to pass freely through it (Fig. 20.8).

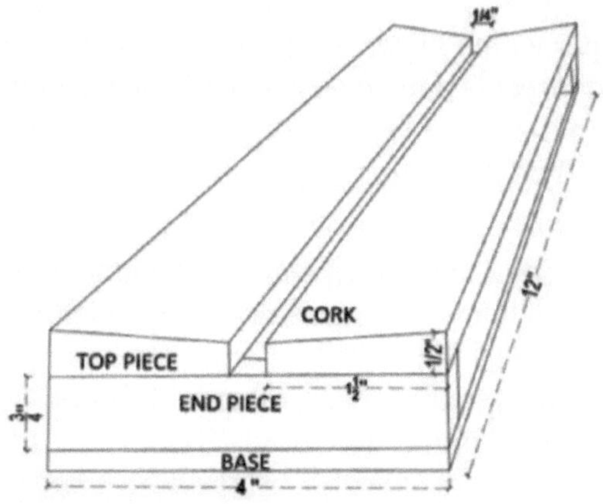

Fig. 20.8 Setting board

For setting of wings where both wings are membranous, they are set with the hind margins of the fore wing at right angles to the body and overlapping the hind wings, so that the pattern of the fore wing can be seen in its full. Where only hind wings are membranous as in Orthoptera, the fore margins of hind wings are arranged at right angles to the body and the forewings are drawn forward to get the wings into position. A paper strip made of card sheet is used across the bases to pin and a broader strip is used to pin across the rest of wing making sure that this is flat and uncreased. Thick pins (No. 2 or No. 3) can be used to flatten the wings.

The antenna may be held under the narrow strips of paper or may be positioned independently and held by pins. Similarly, the legs if they are to be displayed are teased into position with a pin and held there by crossed pins (Fig. 20.9).

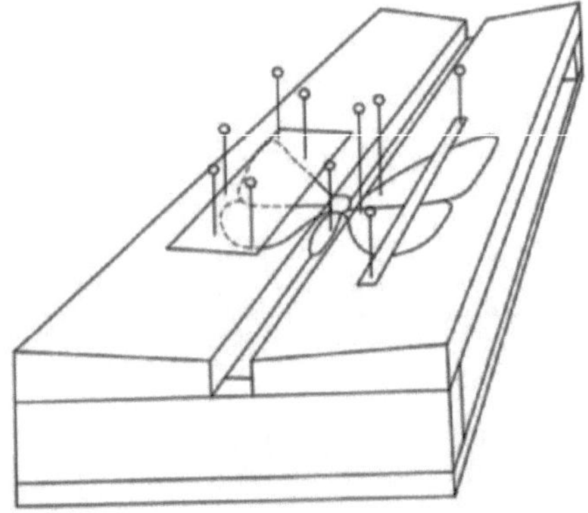

Fig. 20.9 Streching of insect

Upside Down Setting

For bigger insects such as grasshopper which have a large body, upside down setting is recommended. The relaxed insect is pinned upside down on the plain sheet to cork and fixed with paper as before. The legs are stretched out and laid out in the same plane as the wings. When the setting is complete the main pin is removed.

The length of time taken for a specimen to harden depends on the amount of moisture present and on temperature and humidity during drying. Generally, it takes about a week. After drying, remove the pins used for stretching and pull the specimen out from setting board on its mounting pin. Put the label on the mounting pin at once and put the specimen into a store box.

Blowing and Stuffing

Any insect that has large and soft abdomen will shrink badly on drying as well as discoloured by the decay of the internal tissues before drying is completed. In such case abdomen should be emptied as soon as the insect is dead. To empty the abdomen, the insect is placed on a hard surface and a round pencil is placed across the base of the abdomen. Rolling the pencil gently towards the tip of abdomen with moderate pressure cause the intestine to bulge out of the anus. The rolling is to be continued until all the contents of the abdomen have been pushed out. Then abdomen is inflated by blowing in its original shape and sticking the cut portion of anus by mending cement.

Insect larvae particularly caterpillars can be preserved dry by the similar process of rolling and flowing but in this case almost whole body is emptied and more care is needed when re-inflating to recover the true shape. It is advisable to keep larvae alive without food for a day or two so that much waste matter from the intestine may be removed. The larva is then killed in an ethyl acetate killing bottle and immediately rolled starting the pencil just behind the head. When the skin is empty it is inflated and kept inflated during the period of drying in a warm atmosphere.

Blowing apparatus can be prepared easily. A piece of glass tube is drawn out until it is fine enough to be inserted into the anus of larvae. It is clamped in position by a wire clip. The skin can be inflated by mouth but since it has to be maintained under pressure while it is drying up to a period of at least one hour, it is better to arrange some form of pressure reservoir. Rubber tubes with double bulb or an ordinary scent spray bulb can be coupled to a simple reservoir made from toy balloon. In either case, the outer bulb is used to inflate the skin and the reservoir keeps up the pressure until the skin is dry (Fig. 20.10).

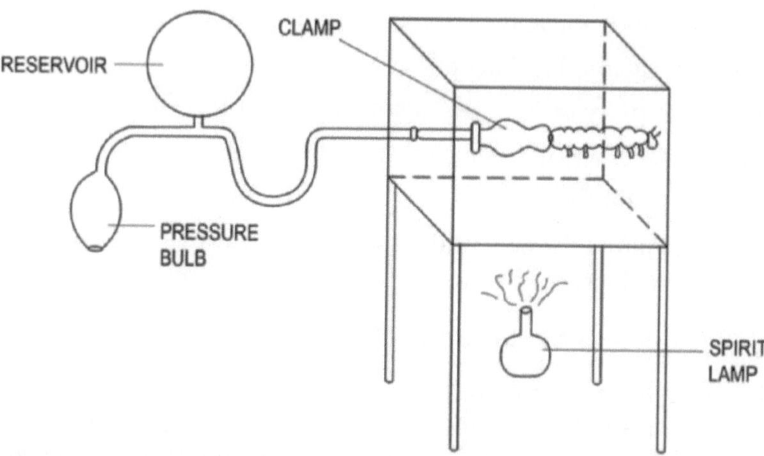

Fig. 20.10 Inflating of a larval skin.

Bulb creates pressure, while it is stored in the reservoir. The spirit lamp provides gentle heat until the skin dries.

Dry Preservation in the Solid State Preservation of Larva

If a soft bodied insect, particularly a larva is to be preserved dry without removing the interior part, it can only be protected against shriveling by removing the moisture either very slowly or very quickly. The very slow method has the advantage that it requires no apparatus but only some reagents. Freshly killed larvae should be used. But in case of dried larvae, they can be softened again by immersing for one or two days in 2% caustic potash solution when they get their natural shape. They must be partially dehydrated by transferring first to 40%, then 70%, 80% and 95% alcohol. Fresh larvae are put straight into 95% alcohol.

Small larvae are kept in 95% alcohol for one week and bigger one for longer period. Then they are kept is absolute alcohol for 3 days which is changed daily. After this the larvae are treated as follows:

1. Transfer to a mixture of one part of xylol to two parts of absolute alcohol.
2. Then transfer in two parts of xylol to one part of absolute alcohol.
3. Finally transfer in pure xylol.

It would be better if they are kept for one day in each mixture. The finished specimen is bottled dry and may be pinned.

Preservation of Specimens

1. Keep in 95% alcohol for 7 days.
2. Then in absolute alcohol for 1 day.
3. Change absolute alcohol, and keep for 1 day.
4. Change absolute alcohol and keep for 1 day.
5. Keep in one part of xylol + 2 parts absolute alcohol for 1 day.
6. Keep in two parts of xylol + one part absolute alcohol for 1 day.
7. Finally keep in xylol for 1 day.

Total 13 days are required for dehydration of specimen. Then dry and pin the specimen. This method has been described by **Van Emden** (1942).

Method of Pinning and Various No of Pins

Insects that can be pinned through the body are those of which the skin is tough enough to grip the sides of the pin, and strong enough to support the weight of the specimen. Very small, fragile or soft bodied insects are damaged if they are pinned, so they must be stuck on a card or celluloid point. Insects of a number of groups especially the small and soft bodied Apterygota,

Thyasnoptera, aphids, lice, fleas and other parasitic group are preserved in liquid and mounted on a microscope slide.

Pinning should be done while the specimen is still fresh. If a hardened specimen is to he pinned, first it should be relaxed. At the time of pinning a suitable pin is used from the start.

Pin and Their Nos

There are 3 series of Entomological pins, English, Continental and "Points" (or minuten).

1. **English pins:** They offer a range, not only in thickness, but also of length, particularly in the range 18-30 mm.

   English pins Nos – 9,10,11,13,16 and 20 are made of stainless steel. They are sold by weight ¼ or ½ and 1 ounce.

2. **Continental pins:** These are used for direct pinning and so they concentrate on length, thinness and sharpness of point. They generally vary in thickness, but available in three lengths:

- 35 mm (Nos. 000,00,0 and 1-7).
- 38 mm (Nos. 8-10).
- 50 mm (Nos. 11-12).
- Nos 2 and 3 are useful size for general use. Common nos. are 2,3,5 and 8 made of steel. They are sold in hundreds.

3. **True minuten:** They are very fine, black pins used for pinning the smallest and softest insects and are extremely fragile in use. they are used exclusively for staging.

Direct Pinning

Direct pinning is done for bigger insects. A pin passes through the specimen and all the various labels, and still has enough length to insert firmly into the bottom of the box. The pin must be sharp as well as long (Fig. 20.11). The drawbacks of direct pinning are:

1. If the specimen is small it is difficult to choose a pin that suits the specimen and get stout enough to hold all the labels firmly and stick into the box without bending.
2. Once the pin has been bent it is nearly impossible to get it truly straight again.
3. At the time of taking off labels to read them, it may easily break the legs.
4. These risks are reduced by following 3 methods of double mounting, this is called Indirect pinning.

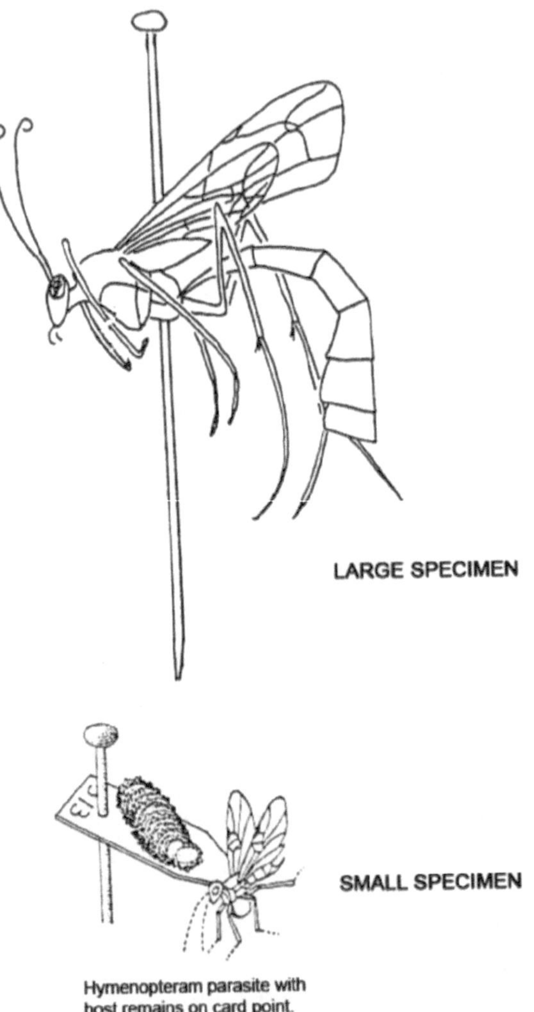

Fig. 20.11 Pinning of large and small specimen

Indirect Pinning

1. Staging

Staging is done for those insects which are small to be pinned directly. They are pinned on to the support or stage. For medium sized insects, a stainless steel, headless pin is used because one with a head is liable to twist suddenly in the pinning forceps and damage the specimen.

For smaller insects, (three) are very fine black steel pins known as "minuten nadeln" or "minuten" (Fig. 20.12).

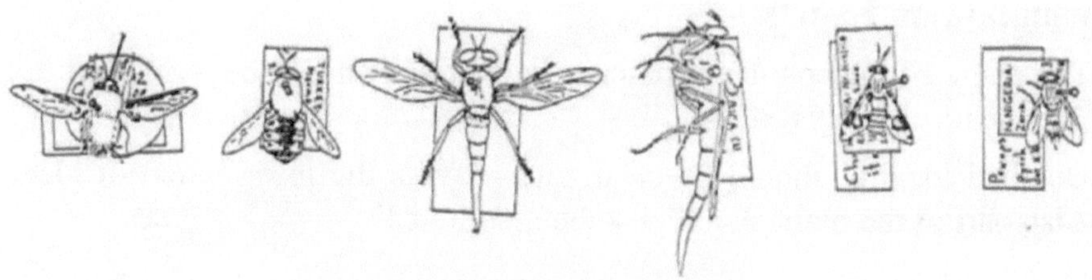

Fig. 20.12 Large and medium sized adults pinned directly through thorax.

2. Carding

A rectangle piece of white card may be used as a stage and the specimen is stuck on it instead of being pinned (Fig. 20.13).

Fig. 20.13 Smaller specimens micropinned and staged on polyporus strips or glued to card points

3. Pointing

This is the best way of mounting the smallest dry insects. A small triangle is cut from the white card. Normally the tip of the point should be a little broader than the thorax of the insect to be mounted on it.

Labelling

A specimen requires certain information for which labels are used (Fig. 20.14). Without labels specimens are useless for taxonomic study.

Fig. 20.14 Small adults and a larva mounted on microscope slides

Information on Labels

1. The name of the country should come first and should be written in full. The locality is essential.
2. Detailed locality should be clear, and provide the information of place. Also part of the plant should also be mentioned.
3. Date of collection.
4. Collector's name.

Example:

India
Uttar Pradesh
CSAUniv. of Agri.& Tech.
Kanpur
On flower
17.01.2021
N. Agrawal

Storage of Insects in Boxes

Permanent collection of pinned insects are kept either in store boxes or in cabinet of drawers.

Store Boxes

These may be either of wood or of card board, but they should be strong. Boxes for permanent storage are invariably rectangular and flat so that they can easily the stacked in a pile or arranged neatly like books on a shelf. Size of box commonly ranges from 10" x 8" to 18" x 12". These boxes may be either single or double sided i.e. they may have specimens in the bottom only or in the lid as well. The pining material is cork, or one of the synthetic material and is covered with white paper. The wood of box should be well seasoned so that it may not turn out of shape. In the front wall of box there should be a small cell for holding naphthalene or paradichlorobenzene. The depth of the box is generally about 1½" on each side, if double sided, including the soft lining material (Fig. 20.15). Extra deep sized boxes are only needed for bigger insects.

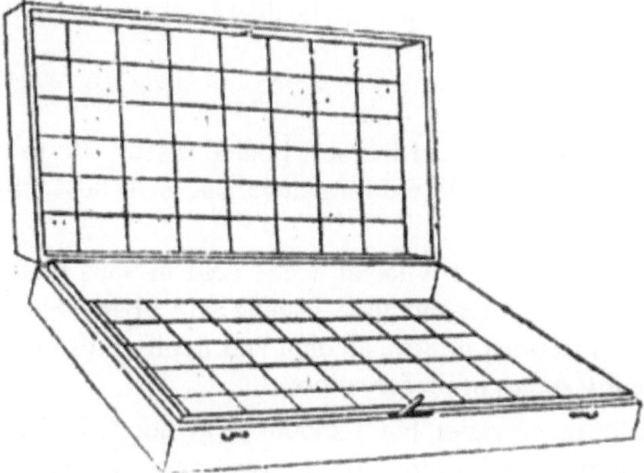

Fig. 20.15 Insect collection box

Cabinets

Cabinets consist of a number of glass topped drawers, each sliding on a pair of runner so that any drawer can be pulled out without disturbing the drawers above and below it. These drawers should be completely interchangeable without sticking or exerting any force.

Labelling of Boxes and Drawers

The labelling of boxes and drawers acts as a guide and an index to the collection. The boxes should be arranged in order, families, genera and species. The outside of box or cabinet drawers should have a general label either stuck on or fitted into a little frame. The label should bear the name of order of insects in capital, the family etc. Inside the drawer rectangular labels should be cut and fixed with short pins.

Protection Against Pests, Mould and Grease

Most cabinet drawers and store boxes have a compartment at the front for preservatives. A piece of gauze or mesh is provided in the wall of the box through which the fumes of the preservative can escape into the interior. The most convenient general preservative is ball of naphthalene. If there is no cell in the box, naphthalene ball should be wrapped in muslin cloth which is tied in each corner of box with the help of pins to avoid loose lumps in the box as they may damage the insects by shaking.

Moulds can be removed by cleaning the specimens in a solution of glacial phenol benzene in the ratio of 1:10 or a dilute formaldehyde.

Grease oozing out or the surface of a specimen can be dissolved by putting the specimen in benzene for few hours.

Liquid Collection

Soft bodied larvae and nymphs of most of the soft bodied insects are to be stored permanently in a liquid. The best liquid for general use is 70-80% ethyl alcohol. Other liquid agents are:

1. Chloral hydrate and Pampel's fluid (Glacial acetic acid in 95% methyl alcohol 12 parts, chloroform 2 parts and glacial acetic acid 1 part). Fix for 12 hours then rinse the specimen in 80% alcohol and finally preserve in 80% alcohol, for the early stages of preservation.
2. Formalin is also used for preservation but it should be avoided as it has a strong handing effect and is troublesome to the eyes and nose when the specimen is being examined under the microscope.

For liquid collection normally glass tubes are used and the size of tube depends on the size of specimens. It is very important to restrict the movement of the specimens otherwise they will soon breakup. These tubes are stored in jars which have the straight side walls, a little higher than the length of the tube with a close fitting lid. The bottom of jar should be covered, either with two layers of blotting papers or shallow layer of cotton wool so that the tubes do not strike against the glass bottom of jar. The alcohol collection should be checked frequently for refilling.

Making Slide Mounts

For making permanent slide mounts we have to do the following steps:

1. **Dechitinisation :** Removal of hardened or chitinised parts of the insect. It will be done by destroying all the soft, internal tissue leaving the rest soft and transparent.

   For this, potassium hydroxide (10% KOH in water) is used both to soften the specimen and to destroy the soft internal tissues. The soft to small specimen may be left in cold potash overnight. Bigger and more substantial specimens may be boiled for about 5 minutes and should have a small cut made in skin underneath the base of the abdomen to allow the potash to penetrate quickly. The next step is to rinse away potash as much as possible with tap water.

   If the specimen is very dark even after washing it may be desirable to bleach it by immersing in a weak solution of bleaching power (NaOCl) or parazone and adding a drop or two of glacial acetic acid.
2. **Staining:** These transparent portion are then stained or bleached as may be necessary.

To stain very transparent specimen such as skin of small larvae, small insects, wings etc., solution of Fuchsine in 2% alcohol is used. Put the specimen in glacial acetic acid and add a few drops of the stain.

3. **Dehydration:** Third step is dehydration of specimen. This is done in small dish like solid watch glass. Put glacial acetic acid in one dish and clove oil in other dish. Transfer the specimen with fine forceps in the glacial acetic acid for dehydration. The specimen is to be put in glacial acetic acid for about 5 minutes to remove all the water from it. Transfer the specimen to clove oil or cedar wood oil for clearing.
4. **Clearing and mounting:** Clearing is done in clove oil. After clearing, display the specimen on glass slide and little Canada Balsam or DPX mountant is dropped on the specimen and cover it with cover slip. It takes few days to harden the mountant therefore, slide must be kept horizontal. The part or specimen is immersed in Canada balsam or DPX on a slide to make a permanent mount.
5. **Labelling:** Finally label the slides (Fig. 20.14).

Direct Mounting

Small soft bodied insects may be mounted direct in one of the compound media such as de Faures, Gum Chloral, Polyvinyl lactophenol or Shellac gel. For these media, no treatment in potassium hydroxide nor dehydrating process is required and the specimen may be transferred direct from alcohol or put into the medium while they are still alive.

21

Entomological Database

A database is a collection of data, typically describing the activities of one or more related organisations. Biological database is meant for storing and organising biological data, collecting information from life sciences, scientific experiments, computational analysis and contents from published literatures.

Biological databases play a vital role in bioinformatics, since it helps varieties of researchers to access and analyse data from different parts of the world. The knowledge obtained helps to solve disease oriented and environment related issues and to make vital policy decisions. Biological databases remain as tool to identify insect species as well as in comparing their relationships with related species. The biological knowledge distributed among different general and specialised databases, can be retrieved with efficient queries.

Various information like genomics, proteomics, metabolomics, microarray, genes expression, phylogenetics etc. are also stored and maintained as biological databases in terms of genome databases, protein structure databases, taxonomic database etc. These computerised web databases facilitate effective data management and analysis help in knowledge transmission, make coherent information exchange between the studies and public audience.

Documenting and preserving the entomological data is an important aspect to fight against insect-transmitted disease in terms of public health and agriculture concern and it is also helpful in formulating the life- saving new chemotherapeutic as well as immunotherapeutic agents. The objective of this scrutiny is to assess the existing entomological database and to identify the major challenges and future perspectives in terms of effective information retrieval. From the observed data, several important inferences are made so that effective counter measures can be taken-in-part. Database management system provides several functions in addition to simple file management.

Database Statistics

Insects are the most diverse of all animals. Until today, over one million insect species were described or identified, which represents more than half of all known living organisms. It has been estimated that nearly six to ten million species may exist worldwide, and potentially they represent over ninety-per cent of different metazoan life forms on Earth. However, less than one per cent insect species are serious pests that affect mankind, livestock and crops.

Though these pests or vectors are very small in number, they are able to cause serious socio-economic, public health and clinical impacts by means of low yield of crops through transmission of several destructive diseases to various crops, humans and animals. In these perspective, understanding about entomology is quiet imperative as insect transmitted diseases impose enormous burden on the world population in terms of loss of life. The humankind suffers due to various vector-borne diseases, particularly in the areas with limited resources therefore, data basing and documentation of insect pests and vectors becomes essential. The taxonomic details of insect species that impact the health of humanity need to be documented or recorded systematically based on phylogenetic relationships of arthropods and related groups, involving identification, classification and naming of organisms.

Taxonomic Database

Taxonomic database is a curated classification and nomenclature for all the organisms in the public sequence database. This currently represents about 10% of the described species of life on planet.

Total number of insect's species identified in the world: 8,61,466.

Total number of insect species identified in India: 58,977.

Largest order: Coleoptera.

Largest family: Curculionidae.

Molecular database on Indian Insects (MODII) is an online database linking several databases like Insect Pest Info, Insect Barcode Information System (IBIn), Insect whole genome sequence, Other Genomic Resources of National Bureau of Agricultural Insect Resources (NBAIR), Whole Genome Sequencing of Honey bee viruses, Insecticide Resistance Gene database and Genomic tools. This database was developed with a holistic approach for collecting information about phenomic and genomic information of agriculturally important insects. This resource database is available online at http://cib.res.in.

Insect Barcode Information (IBIn)

The Consortium for the Barcode Life (CBOL), International Barcode of Life (IBOL) and European Barcode of Life (EBOL) are devoted for the global standard for the identification of biological species by developing database. The database developed by the National Bureau of Agricultural Insect Resources (NBAIR), Bangaluru would be helpful to all entomologists and others in the country as ideal platform for their genomics work.

The Insect Barcode Information (IBIn) is a platform to assist and manage acquisition, storage, analysis and to explore DNA barcode records for species identification and genetic analysis of status data of Indian insect species and their resources along with GPS data. IBIn is comprised of interspecific and intraspecific records of genera collected from different parts of India. The web version of information incorporates browser interface facilitating statistics, taxonomy, application species identification and sequence diversity estimation. An attention is to be made to increase the database content with the additional curated and computed data on the barcode process as well as expand the analytical capabilities of the web interface with further novel data mining and visualisation tools.

Insect Barcode Database

Total number of species barcoded in the world: 1,10,623.

Total number of species barcoded in India: 2008.

Table 3: Barcoded species in different insect orders

S.No.	Order	Barcoded Species in The World	Barcoded Species in India
1.	Lepidoptera	62934	1227
2.	Hymenoptera	16026	317
3.	Diptera	8905	178
4.	Coleoptera	8115	109
5.	Hemiptera	3409	110
6.	Orthoptera	804	1
7.	Odonata	359	15
8.	Neuroptera	142	2
9.	Blattaria	125	0
10.	Mantodea	150	0
11.	Ephemeroptera	608	1
12.	Isoptera	197	0
13.	Megaloptera	27	0
14.	Mecoptera	27	0
15.	Dermaptera	11	0
16.	Embioptera	11	2
17.	Diplura	4	0
18.	Grylloblattodea	1	0
19.	Mantophasmatodea	1	0

Colour used in Insects Barcoding

Adenosine- Green (Purine).

Thiamine- Red (Pyrimidine).

Cytosine- Blue (Purine).

Insect Molecular Database

The bioinformatics data analysis methods have been of prime importance to enable the scientists in getting the new information about the insects. Genome sequencing focuses mainly on genomics and proteomics projects on multiple insects. Therefore, the bioinformatics researchers and professionals have also commenced the development of insect customised databases and analysis tools. The molecular biologists are using high throughput genomics, transcriptomics, regulatory genomics and proteomics methods to decipher the system level biology, physiology, disease mechanisms, host-pathogen interaction, insect resistance, growth and development of various important insects. This will enable a detailed survey of insect specific data resources and application tools on genomics, transcriptomics, proteomics, regulatory genomics, miRNA profiling, RNA interference studies etc. of all important insects.

Molecular database mainly categorizes three sections which give both molecular and morphological information about some important insects:

1. Genomics.
2. Proteomics.
3. Transcriptomics.

1. Genomics

It is a tool that can be applied to facilitate research in a wide variety of areas, and the availability of insect genomes. There are almost 230 insect genome sequencing projects initiated by different Institutes, universities and research centers. Among these, only 14 genomes are complete and 94 genomes are in continued phases.

2. Proteomics

It is a very powerful technique that can be applied to interrogate changes at the protein level. It provides a mechanism for diagnosis and therapy of infection, diseases, response to drugs and biological aging or gradual deterioration of functional characteristics.

3. Transcriptomics

Study of the complete set of RNAs encoded by the genome of a specific cell or organism is known as Transcriptomics. It plays an important role in knowing the quantity of specific proteins which are responsible for the specific mechanism of an insect.

Database Management System Design

Database design is the process of producing a detailed data model of a database and it stands for the overall process of designing, not just the base data structure, but also the forms and queries used as part of the overall database application within the database management system. Biological databases and new data models are sensitive to the novel characteristics of biological data.

Data Processing and Analysis

Each collected data is cleaned, checked for completeness, coded and analysed with an IBM compatible microcomputer using Statistical Package for Social Science (SPSS). Data integration is highly necessitated with increasing frequency as the volume and the need to share existing data explodes.

Specific Database

Species specific databases are available for some species. Nearly 18% of entomological databases are observed as species specific i.e. some dipterans and coleopterans.

Data retrieval

Extracting required information from the database by querying is known as data retrieval. There are two types of querying:

- User can specify the query for the data to be retrieved.
- Automatic query builder that reduces the burden of user.

The observed entomological databases reveal that no database is there to have an efficient way of adaptive querying in order to provide the biologists with ease of usage, like providing tools to modify or optimise queries based on previous search etc. Therefore, providing intelligent interfaces for structural entomological databases shall help to obtain fast, effective and optimised results.

Challenges and Opportunities in Entomological Databases

There are several challenges with the entomological databases that are relative to any other biological databases, retrieval of information, complex querying, extension of database, integration of databases, in providing clarifications and versioning facility and in authentication. Similar to other biological databases, sustainability issue is observed with many websites that become inaccessible. Maintenance of digital data in long term basis in the databases has also faced certain challenges.

Aggregation of information stored in different databases is required to be maintained on long term basis by major institutions and government agencies, since maintenance in long term basis is expensive.

Entomological Database Resources

Websites

www.nbair.res.in

www.cib.res.in

www.entsoc.org

www.insectidentification.org

www.amentsoc.org

www.ndsu.edu

www.csiro.au

www.ent.istate.edu

Journals

Indian Journal of Entomology (India)

Journal of Insect Science (India)

Entomon (India)

Indian Journal of Plant Protection (India)

Biocontrol (formerly *Entomophaga*), IOBC, Paris

International Journal of Entomological Research

Journal of Entomological and Zoological Studies

Applied Entomology and Zoology (Japan)

Annual Review of Entomology (USA)

Museums

IARI, (Pusa Institute), New Delhi

TNAU, Coimbatore

British Museum (Natural History), London, U.K.

Swedish Museum of Natural History, Stockholm, Sweden

American Museum of Natural History, New York, U.S.A.

References

Agrawal, Neerja (2021). Fundamental Entomology: A Practical Manual. Jaya Publishing House, Delhi. pp.108-146.

Agrawal, Neerja and Pandey, N.D. (2022) Objective Entomology. Kalyani Publisher, New Delhi. pp.3-24.

Boucek, Z. (1986). Nomenclature. Handout, 8th International Course on Applied Taxonomy of Insects and Mites of Agricultural Importance. C.I.E. London, 1986. pp. 1-6.

Imms, A.D. (1925). A General Text Book of Entomology. (Revised by Richards and Davies 1957), Methuen & Co. Ltd. London.

Kapoor, V.C. (2008). Theory and Practice of Animal Taxonomy. Oxford and IBH Publishing Co. Pvt. Ltd. New Delhi pp. 71-82.

Mayr, E. (1969). Principles of Systematic Zoology. Hill Publishing Company Ltd., New York, pp. 121-297.

Mayr, E.; Linsley, E.G. and Usinger, R.L. (1953). Methods and Principles of Systematic Zoology. McGraw- Hill Book Company, INC. pp. 212-279.

Oldroyd, Harold (1970). Collecting, Preserving and Studying Insects. Hutchinson and co., London.

Johansen, C. (1978). Classification of Insects and their relatives in Fundamentals of Applied Entomology 3rd ed. (ed. Pfadt), MacMillan. pp.108-149.

Sehgal, V.K. (1999) Insect biosystematics and sustainability in agriculture. Tenth Dr. C.P. Alexander memorial Lecture. Department of Zoology, University of Delhi. pp.1-16.

Sockal, W.H. and Sneath, R.R. (1963). Principles of Numerical Taxonomy. W.H. Freeman and Company, California.

White, I.M. (1983). Modern Classificatory analysis. Handout, 8th International Course on Applied Taxonomy of Insects and Mites of Agricultural Importance. C.I.E. London, 1986. pp. 1-5.

Glossary

Systematics: Systematics is the scientific study of the kinds and diversity of organisms and any or all relationships among them. The term coined by a Swedish naturalist Linnaeus (*Systema Naturae* 1735), is a Latinized Greek word *Systema*, as applied to the systems of classification.

Taxonomy: Taxonomy is theoretical study of classification, including its basis, principles, procedures and rules. The name taxonomy was first proposed by de Candolle (1813) for the classification of plants. The term taxonomy is derived from a Greek word – *Taxis* means arrangement and *nomy* means law. Thus arranging the living organisms according to certain law is taxonomy.

Classification: Ordering of organisms into groups (or sets) on the basis of their relationships that are of associations by contiguity, similarity or both.

Nomenclature: *No-men-cla-ture* means a system of names. It is the application of distinctive names to different taxa. It is derived from Latin *Nomen* means name and *calor* means to call, which means to call by names.

Binomial nomenclature: Linnaeus in his 10th edition of Systema Naturae (1758), used double name uniformly and constantly for all kinds of plants and animals. First they are grouped to have an idea of interrelationships of living organisms under genus. All insects which possess well defined and constant character of form and structure which interbreed are grouped under a species.

Trinomial Nomenclature: Some insects have trinomial nomenclature. These names are geographical races, which are called "varieties".

Phylum: A large group of taxa (singular Taxon, which includes many classes).

Class: A unit of Classification in animal kingdom which is a part of the phylum and includes many orders.

Family: A classification category that includes a number of genera sharing one or a number of characteristics, ending in suffix "idae"(eg.Acrididae).

Genus: A special group or taxa of animals having similar characters which include many species. In the Binomial Nomenclature, the first name is genus which starts with capital letter.

Species: Groups of actually (or potentially) interbreeding natural populations which are reproductively isolated from other such groups.

Types: Whenever a new species or other group is described, a describer is supposed to designate a type, which is used as a reference, if there is ever any question what that species or group includes . The type of a species or subspecies is a specimen, the type of a genus or subgenus is a species and the type of a family or sub family is a genus.

Biotype: A population or group of individuals composed of a single genotype .

Genotype: In nomenclature, the type species of a genus (ef. Type species); in genetics, the class in which an individual falls on the basis of its genetic constitution, without regard to visible characters (ef.Phenotype).

Strain: A biological strain of an organism, morphologically indistinguishable from other members of its species but exhibiting distinctive physiological characteristics; particularly in regard to its ability to successfully utilize pest-resistant host organisms or to act as an effective beneficial species.

Holotype: Species selected by the author of the species from the type species.

Syntype: A series of specimens designated by the author of first species at the time of establishing species.

Lectotype: A specimen selected by reviser from among Syntypes.

Neotype: In case Holo, Syn or Lecto types are lost or destroyed, a new specimen is designated.

Specific name: "The binominal combination of a generic name and a specific trivial name which constitutes the scientific designation of a species" (International Commission 1948); also used by many workers and in the original rules in place of trivial name.

Scientific name: The binomial or trinomial designation of an animal, the formal nomenclatural designation of a taxonomic category.

Senior homonym: The earliest published of two or more identical names for the same or different taxonomic categories.

Senior Synonym: The earliest published of two or more available synonyms for the same taxonomic unit.

Primary homonym: One of two or more identical trivial names which, at the time of original publication were proposed in combination with the same (or an identical) generic name (eg- *X-us albus* Amith 1910 and *X-us albus* Jones 1920); the later of such primary homonyms are to be permanently rejected; also one of two or more identical names for genera or higher categories.

Secondary homonym: One of two or more identical trivial names which, at the time of original publication ,were proposed in combination with different generic names but which through subsequent transference, reclassification or combination of genera have come to bear the same (or an identical combination) trivial name. Bulletin of Zoological Nomenclature, 4:97-105 (1950).

Phylogeny: The study of the historical development of the line or lines of evolution in a group of organisms; the origin and evolution of higher categories.

Polytopic: Occurring in different places as, for instance, a subspecies composed of widely separated populations.

Polytypic: A category containing two or more immediately subordinate categories, as a genus with several species or a species with several subspecies.

Natural classification: As currently used, classification based on characters or groups of characters which indicate phylogenetic relationship.

Law of Priority: The provision in the international rules of zoological nomenclature that the correct name for a genus or species can be only that name under which it was first designated in conformance with the requirements laid down in these rules.

Hierarchy: In classification, the system of ranks which indicates the taxonomic level of various taxonomic categories (i.e. kingdom to species).

Sibling species: This name is applied to pairs or groups of very similar and closely related species. They occur commonly from protozoa to mammals. Sibling species are not a separate taxonomic category. They do not differ from other species in any respect for the miniatures of their structural differences.

Polytypic species: It was found that some species are wide spread and consist of many local populations. If these local populations are sufficiently distinct from each other they are called subspecies. Species that consists two or more subspecies are called polytypic species.

Monotypic species: Species which have no subspecies or to be more precise, consist of only a single subspecies are called monotypic species.

Super species: Super species is a monophyletic group of very closely related and largely or entirely allopatric species.

Subspecies: Subspecies are geographically defined aggregate of local populations which differ taxonomically from other such subdivisions of a species.

Not more than one subspecies can exist in breeding condition in any one area. Adjacent subspecies interbreed or are potentially capable of doing so if separated by extrinsic barriers.

Race: Subgroup or biotype within a species or variety, which are morphologically identical but differs from other races by virulence, symptoms or host range.

Variety: A loosely defined taxonomic category below the level of sub species, usually possessing some visibly different characters.

Category: Rank or level in a hierarchy into which natural populations are classified, such as subspecies, species, genus and family.

Taxon (pl. taxa): Taxonomic unit or rank.

Trivial name: The second or third word in a binominal or trinominal name of an organism; the specific and subspecific components in the scientific designation of an animal, such as Specific trivial name, Subspecific trivial name, Infrasubspecific trivial name.

Entomological Terms

Abiotic factors: Non-living or environmental factors that influence the rate of multiplication of any organism. They include topography, relative humidity, temperature, light, wind, soil etc.

Biotic factors: The relationship among biological organisms when they live together, influence each other's life directly or indirectly for growth, nutrition and reproduction. These are living factors called biotic factors. They include 1. Relationships among organisms, 2. Interactions with plants, 3. Food in relation to insects.

Acaricide: A chemical or chemical formulation which is used to kill mites. Also called miticide.

Acephalus larva: Larva with no head capsule or reduced head appendages. It is also called maggot e.g. larva of house fly and many dipterans.

Active ingredient: Toxic component present in a pesticide.

Aestivation: Dormancy during summers or dry seasons. Also called "Summer sleep".

Aflatoxins: Organic compounds or metabolites produced by fungus *Aspergillus flavus*, which are highly toxic and carcinogenic to mammals.

Alate: Winged forms in the life cycle of certain insects such as aphids.

Amino acids: The basic organic compounds that contain the amino group and the carboxyl group required for the synthesis of proteins.

Annules: A ring like marking or a ring of hard cuticle.

Antifeedent: A natural or chemical substance, which acts either to inhibit the stimulation of gustatory receptors to recognize suitable food or stimulate the receptors to elicit a negative response to food.

Apodous larva: Degenerate type of larva common in weevils, wasps, bees and flies, in which body appendages are completely supressed and legs are absent e.g. larvae of dipteran flies.

Apterous: Wingless forms in the life cycle of certain insects such as aphids.

Attractants: Any material that is able to attract insects and make them eat or contact the poison kept in baits to kill them.

Balancers or halteres: Reduced hind wing shaped drumsticks in Diptera. These balancers constantly vibrate in flight so that any change in altitude by the fly causes an activation of tension receptors.

Brachypterous: Winged adult insects such as aphids. Also, a condition of short wings which do not cover the entire abdomen e.g. brown plant hopper.

Broad spectrum insecticides: Insecticides effective against a variety of pests or a broad range of targets, generally acting on insects' nervous system; non selective and having same range of toxicity to a wide range of insects.

Campodeiform larva: Larva more or less fusiform with depressed body, well sclerotized prognathous head, long thoracic legs and a pair of cerci. They are the most primitive form of larva and have no compound eyes or ocelli. Example, Neuroptera, some Coleoptera, Strepsiptera and Trichoptera.

Carabiform larva: Larva shaped like the larva of a carabid beetle, that is elongate, flattened and with well-developed legs. Filaments lacking at the end of abdomen.

Caterpillar: Polypod or eruciform larva is called caterpillar e.g. Lepidoptera and saw fly.

Crysalis: A naked pupa; an open pupa, not enclosed or covered; e.g. pupa of butterfly.

Cocoon: A silken case woven by a full grown larva or a fibrous case made by a larva or an earthen case in the soil made by a larva, in which it pupates.

Colleteral hosts: Host plants belonging to species other than the species of the primary host, but within the same genus.

Crawlers: The early stage nymphs of mealy bug, which have functional legs and can crawl about short distances.

Defoliator: Any biting and chewing insect that destroys the leaves of plants extensively.

Deterrent: A chemical substance that deters feeding or oviposition of an insect.

Deutonymph: The third instar stage in the development of mites and ticks.

Dioceous: Having the male and female sex organs in different individuals. The individuals may be either male or female.

Diurnal: Active only during daytime or habitually flying in day only.

Dormancy: A state of quiescence or inactive stage.

Ecdysis or moulting: The process of shedding the skin (exoskeleton) which occurs periodically during the development of insect or other arthropods.

Ecology: The study of living organism in relation to its environment with other organisms.

Economic injury level (EIL): Infestation level of a particular pest at which the damage caused by the pest will result in economic loss. It is the lowest population density of a pest that causes economic damage.

Economic threshold level (ETL): Pest population level at which control measures have to be initiated to prevent the pest population reaching the economic injury level. It is the infestation level of a particular pest beyond which there will be economic loss.

Ecosystem: A self-contained habitat in which living organisms and the environment interact in an exchange of energy making a continuing cycle.

Ectoparasite: A parasite that live externally and feeds from the host by its specialised mouth parts or organs to take nourishment from the host. It feeds externally and does not kill the host.

Egg parasitoid: Parasitoid that deposits their eggs inside or outside of host eggs. The progeny of the parasitoid emerging from the host eggs kills the eggs. Example- *Trichogramma* spp. on lepidopteran eggs.

Endemic pest: A pest permanently established in a moderate or severe form in a defined area, commonly a country or part of a country.

Endoparasite: A parasite which enters and lives within the host and feeds from the host.

Entomophagous: Organisms which feed on insects or their parts; insectivorous.

Entognathus: The mandibles and maxillae remain retracted into pouches in the head. Example- Protura, Collembola and Diplura.

Ephemeral: Short lived.

Epidemic (Epiphytotic): A sudden flare up of a pest from its endemic area which is widespread and severe. It may be a temporary flare up in a particular season.

Eruciform larva (Polypod larva): A larva with a more or less cylindrical body, a well-developed head and with both thoracic legs and abdominal prolegs e.g caterpillars of moths and butterflies.

Eucephalous larva: Larva with a distinct well developed sclerotized head capsule, mouth parts and antennae. Example- larva of mosquito.

Exarate pupa (Dectitus pupa): They are also called free pupa or pupa libra. Pupa remains uncovered, without puparium. All the appendages such as wings; legs are free from the rest of the body. The pupa resembles adult e.g. grub of beetles.

Exuviae: The cast off skin of nymph or larvae at the time of metamorphosis.

Fecundity: The number of eggs produced regardless of fertility.

Gregarious: Forming aggregation.

Grub: Larva of Coleoptera and Hymenoptera.

Hemelytra: Wings in which basal half is thick and leathery while the distal half is membranous e.g. forewing of heteropteran bug.

Hemicephalus larva: The head capsule and appendages are considerably reduced and the head is retracted into the thorax. Head capsule is considered between eucephalous and acephalous larva. Example- Brachycera, Tipulidae.

Hibernation: State of inactivity during winter. Also called winter sleep.

Hypognathous: Having the head vertical and mouth directed downwards, e.g. Orthoptera.

Imago: Adult or sexually matured form of an insect.

Instar: Stage between moults.

Integrated control: A pest or disease control approach that uses all available method of control, which is economical and with least damage to the environment.

Intermediate host: Host that harbours immature stages or asexual stages of a parasite, e.g. mosquito for the malarial parasite.

Juvenile hormone (JH): A hormone secreted by corpora allata into the haemolymph that maintains the immature form of an insect during early moults. It consists of a multi carbon chain.

Juvenile stage: Every post embryonic stage in the development of an organism that leads to sexually mature stage such as, egg larval and pupal stages of insects.

Key pest: An important major pest species in the pest complex attacking a crop and causing economic damage, with a dominating effect on control practices.

Knockdown effect: An insecticide that has power to knockdown insects rapidly but may not kill them immediately.

Larva aculeata: Larva with dense, fur like hairs.

Larva cornuta: Larva with fleshy horns or processes.

Larva furcifera: Larva with furcate processes.

Larval parasitoids: Parasitoid that deposits its eggs in or on the host larvae. The progeny of the parasitoid emerges from the host larvae, parasitizes and destroys the host larvae, e.g. *Bracon brevicornis*, *Apanteles flavipes* etc.

LD_{50} or Median lethal dose: A value used in presenting mammalian toxicity of a toxicant, usually oral toxicity, expressed as mg per kg of body weight; a lethal dose for 50 per cent of the test organism or 50% mortality in a pest population.

Longevity: Life span of an individual.

Maggot: Larva of Diptera, vermiform, devoid of legs and without a well-developed, distinct head capsule.

Major pest: Insect pest that causes a loss of more than 10 per cent of the produce.

Malformation: Part of a plant that is deformed due to some causing organism.

Matrix: The gelatinous substance into which eggs of *Meloidogyne* and some other nematodes are deposited to form an egg mass.

Mechanical control: Any mechanical pest control device or means such as insect and mammalian traps, barriers to prevent entry or gaining access to plants or other commodities.

Metabolism: The entire process or series of changes, or transformation of ingested food into tissues and cells to produce energy and later convert into waste products.

Metamorphosis: The series of changes which most of the insects undergo as they develop into adult.

Methyl eugenol: A chemical sex attractant for fruit fly *Bactrocera* spp.

Migratory: Migrating from one place to another in search of food and shelter.

Minor pest: Insect or non- insect pests that cause a loss ranging from 5- 10%.

Naid: Aquatic nymph such as of dragon fly, mayfly or other hemimetabolous insects.

Nematicide: A chemical compound or physical agent that kills nematodes.

Nematode: Worm like, generally microscopic animals that live saprophytically in water or soil or as parasites of plants and animals.

Nocturnal: Active during night.

Nymph: An immature stage which is similar to the adult except for the absence of functional wings and sexual organs.

Obligate parasite: A parasite that can lead a parasitic mode of life and can not live in any other way.

Obtect pupa: The pupa in which the appendages and body are compactly fused by a hardening of the outer shell or theca, e.g. moth pupa.

Occasional pest: Pest that reaches significant levels only occasionally, e.g. rice case worm.

Oligopod larva: Larva has a well- marked head thorax and abdomen. Thorax bears three pairs of legs but no abdominal legs. Example- larva of beetles.

Omnivorous: Feeding on any kinds of animal or plant food.

Opisthognathous: Having the front of head turned backwards so that the mouth parts project to the rear, e.g. Hemiptera.

Oviparous: Reproduction by eggs laid by female.

Parasite: An organism deriving all or portion of its nutrition from another organism (host), either permanently or temporarily.

Parthenogenasis: Development of eggs without being fertilised; reproduction from unfertilised eggs.

Permanent pest or persistent pest: Pests which always remain on or near the host whether they have a long or short life cycle. They may have a resting period either on host or in the soil, e.g. mango hopper, coccids etc.

Pest: Term pest is derived from Latin word "Pestis" which means causing destruction. Pests are defined as those organisms, which compete with man in his food supply, damage his possessions and attack his persons. Pests include insects, mites, ticks, nematodes, fungi, bacteria, weeds, rodents, birds molluscs crustaceans etc.

Pest resurgence: Rapid increase in numbers of pest following cessation of control or resulting from development of resistance of a particular pest to certain pesticide or due to elimination of its natural enemies as a result of pesticide application.

Pheromone: Chemical substance that attracts members of the same species or one sex of that species, especially insects and nematodes.

Physical control: Control of pests with physical means such as, heat, cold, electricity, sound wave etc.

Phytophagous: Herbivorous; plant eating.

Phytotoxicity: Substances causing adverse effect to plants, may be toxic to them. These include pesticides, fungicides, nematicides etc.

Poison bait: An attractant food stuff for insects, rodents and molluscs which when mixed with a toxicants substance to attract and kill the target organism.

Polyphagous pest: Pests infesting more than one host species.

Polypod larva (Eruciform larva): Larva has a well-developed head, three segmented thorax, each bearing a pair of clawed legs and an abdomen of ten segments bearing five pairs of fleshy, hooked legs called prolegs. Example- caterpillar of a butterfly.

Predator: An organism eating and living on other organisms.

Prolegs: The abdominal legs of caterpillars (Lepidoptera, saw flies), bearing a characteristic arrangement of crochets; also called pseudo legs, false legs, prop legs or abdominal legs.

Pronotum: First segment of dorsal surface of thorax.

Prothoracic shield: Thick hard protective covering of first segment of thorax.

Protonymph: The first nymphal stage of a tetranychid mite. It is also called larval stage.

Pseudocoel: A false coeloms or body cavity not lined with an epithelium of mesodermal origin, containing a liquid with various internal organs.

Pterygotes: Winged insects.

Pulverulent: Powdery; dusty; covered with very minute powder like scales.

Pupa: A resting stage in insect metamorphosis between larva and young adult, during which movement and feeding stops.

Pupa adecticous (pupa complex): The appendages remain in separate sheaths and only partly free. Example- mosquito pupa or 'tumbler'; or adhere to the body and the abdomen is capable of movement, e.g. most lepidopteran and dipteran pupa; or enclosed in the last larval skin or puparium, e.g. some dipteran pupa.

Pupa coarctata: The last larval skin is not cast off but formed into a puparium. The insect inside cannot be seen. The adult insect emerges from the puparium by making a circular opening on the pupal skin. Example- pupa of house fly, stable fly etc.

Puparium (Pupal case): After the last moult of the larva, the larval skin is not cast off, but hardens and become chitinous to form a case or covering.

Pupa obtecta: The legs and wings are bound down to the body by moulting fluid and can be seen externally while the pupa is active. Adult insect emerges from the pupal case through a T shaped slit on its dorsal surface, e.g. mosquito, sand fly and horse fly.

If the appendages are tightly oppressed to the body, covered by an external skin and abdomen is capable of movement, such pupa is called chrysalis. It has a posterior process or cremester, which anchors with the help of apical hooks such as some lepidopteran pupa.

Pupa nude: A naked pupa which is free of any attachment.

Quarentine: All operations associated with the prevention of importation of unwanted organisms into a territory or their importation from it. These regulations are imposed usually to prevent a disease, insect, nematode or weed invasion of an area.

Quiscent: A stage of biologically inactive phase induced by physical factors such as, low temperature or other adverse environmental conditions.

Random pests: Casual visitors such as weevils, cockchafer beetles, blister beetles which have no seasonal occurrence.

Reproductive potential: The maximum possible rate of reproduction of an organism. This is also called as 'reproductive capacity' or 'biotic potential'.

Rhizosphere: The soil immediately surrounding the roots of living plants; area in the soil immediately surrounding plant roots and influenced by them.

Scarabeiform larva: A grub like, stout, sub cylindrical, C shaped sluggish larva with a well developed head, short thoracic legs but no abdominal legs and has soft fleshy body, e.g. white grub beetle.

Scutellum: The third dorsal sclerite of the mesothorex, a triangular piece between the elytra of Coleoptera, the triangular part of mesothorax between the bases of hemelytra; a hemispherical part cut off by an impression of line from the mesonotum in Diptera.

Sedentary: Stationary without any body movement; remaining in a fixed position.

Seasonal pest: Pests that occur during a particular season on a specific crop, e.g. red hairy caterpillar on groundnut.

Semilooper: A caterpillar in which only one or two pairs of legs are wanting and when they move, small loops are formed, e.g. castor semilooper.

Sessile: Permanently attached.

Seta: A slender hair like growth of the integument.

Severe pest: A pest with a general equilibrium position above the economic injury level, which makes the pest a constant problem.

Sex pheromone: A substance generally released by the female to attract males for the purpose of mating.

Social insects: Insects that live in groups making colonies with different casts or forms for specific functions in the colony. Example- termites, honey bees etc.

Sporadic pest: Pests that infest certain crops occasionally within some limited areas or pockets in a more or less severe form. Example- rice earhead bug.

Stadium: The time interval between each moult in a developing insect.

Stylet: Stiff, generally long slender, hollow, tube like feeding organ of sap sucking insects like aphids, leafhoppers, plant parasitic nematodes etc.

Taxis: A direct response involving the movement of a living organism towards or away from a stimulus.

Termitarium: The earthen structure or nest inhabited by a termite colony.

Tibial spur: A large spine on the tibia usually located at the distal end or tibia.

Tracheal gills: Larval stages of aquatic insect breath under water through gills. These are outgrowths of tracheae; or respiratory organs of some aquatic insects consisting of outgrowths of the body wall containing numerous enclosed tracheoles in which dissolved oxygen can diffuse from the surrounding water.

Trail making pheromone: An intraspecific chemical messenger produced by foraging ants and termites to indicate source of provisions to other colony member.

Vector: A living organism able to carry or transmit a pathogen, especially insect, nematodes etc.

Vermiform larva: Relatively long and slender larva, worm shaped, typical of some Diptera.

Vestigial: Having a nature of degenerate or atrophied organ, that was fully functional in an earlier stage of development of the individual or species.

Virulence: State of being pathogenic.

Viviparity: A type of reproduction in some species of insects in which the embryonic development is completed within the body of the female parent and therefore, instead of eggs, young larvae or nymphs are laid.

Weevils: Insects belonging to Order Coleoptera with prominent snout in adults.

Wrigglers: The aquatic larval stage of mosquitoes.

Zygote: A fertilised egg.

Expected Questions

1. In which order of insects do cerci are found?
2. What is the position of head in Mantoidea?
3. To which family does silkworm belong to?
4. The hind wings in Dipterans are represented by a pair of?
5. What is the full form of CAB?
6. What is the book name written by Linnaeus on Taxonomy and in which year it was published?
7. In which order forewings are modified into leathery elytra?
8. What is the lowest category in classification?
9. Who introduced the term binomial nomenclature?
10. Who studied evolution of animals?
11. Who wrote the book "Principles of systematic zoology"?
12. Which order do dragonflies belong to?
13. To which family do *Holotricha consanguinea* belong to?
14. To which order and family Robber fly belongs to?
15. Which super family does mulberry silk moth belong to?
16. Which order do bristle tails belong to?
17. Which order do braconid belong to?
18. Which is the lowest category in classification?
19. Which is the smallest insect order that has been reported in 2002?
20. How many suborders do order Lepidoptera divided into? Name them.
21. Name the Hemipteran insect and family which secrete sweet substance?
22. Give name of insect order represented by the following taxa?

    a) Cecidomyidae   b) Noctuidae
23. Exarate type of pupa is common in which order?

24. To which family do ants belong to?
25. What is the highest category in classification?
26. Chalcid parasite belongs to which order and family?
27. To which order does sliver fish belong to?
28. In which order all the thoracic and abdominal ganglia have coalesced to form a single mass?
29. Which class do millipedes belong to?
30. Insects belong to which subphylum?
31. Hemelytra type of wing modification is found in which suborder?
32. Pyrilla belongs to which family?
33. Anal horns in the larval stage are found in which family?
34. Body in Crustacia is divided into which parts?
35. The study of the kinds and diversity of organisms and the relationship among them is known as?
36. The theory and practice of identifying, describing, naming and classifying organisms is known as?
37. A group of inter-breeding natural populations which are reproductively isolated from other such groups are known as?
38. The arrangement of organisms into groups on the basis of their relationship is called as?
39. Name the taxa between which cohort is present.
40. Accordng to Intenational Code of Zoologicl Nomenclature tribe ends with which suffix?
41. When one name is given to two or more taxa, what is it called?
42. When different names are given to same taxa what is it called?
43. What is the connecting link between Apterygota and Pterygota?
44. Click beetles belong to which family?
45. Presence of pearman's coxa which acts as stridulatory organ is seen in which order?
46. Bubenic plague is transmitted by which insect?
47. In which group of Lepidoptera mouthparts are well developed with the functional mandibles?
48. Veinless wing bearing hymenopteran parasites belong to which family?

49. The word Arthropoda has been derived from which language?
50. In which order antennae are absent?

Answers

1. Dictyoptera
2. Hypognathous
3. Bomycidae
4. Haltere
5. Commonwealth Agricultural Bureau
6. *Systema naturae*, 1735
7. Coleoptera
8. Species
9. Carolus Linnaeus
10. Darwin
11. E. Mayr
12. Odonata
13. Scarabaeidae
14. Diptera, Asilidae
15. Bombicoidea
16. Thysanura
17. Hymenoptera
18. Species
19. Mantophasmatodea
20. (i) Monotrysia (ii) Ditrysia
21. Aphid, Aphididae
22. (a) Diptera (b) Lepidoptera
23. Coleoptera
24. Formicidae
25. Kingdom
26. Hymenoptera, Chalcididae
27. Thysanura
28. Hemiptera

29. Diplopoda
30. Uniramia
31. Heteroptera
32. Fulgoridae
33. Bombycidae
34. Cephalothorax and abdomen
35. Systematics
36. Taxonomy
37. Species
38. Classification
39. Class and order
40. ini
41. Homonym
42. Synonyms
43. Lepismatidae of Thysanura
44. Elateridae
45. Psocoptera
46. Rat flea
47. Micropterygidae
48. Platygasteridae
49. Greek
50. Protura

Index

J

K

L

Q

R